Collins

Teacher Pack 2.2

Maths Frameworking

3rd edition

Rob Ellis, Kevin Evans,
Keith Gordon, Chris Pearce,
Trevor Senior, Brian Speed,
Sandra Wharton

William Collins's dream of knowledge for all began with the publication of his first book in 1819. A self-educated mill worker, he not only enriched millions of lives, but also founded a flourishing publishing house. Today, staying true to this spirit, Collins books are packed with inspiration, innovation and practical expertise. They place you at the centre of a world of possibility and give you exactly what you need to explore it.

Collins. Freedom to teach.

Published by Collins
An imprint of HarperCollins*Publishers*
77–85 Fulham Palace Road
Hammersmith
London
W6 8JB

Browse the complete Collins catalogue at
www.collins.co.uk

10 9 8 7 6 5 4 3 2 1

ISBN-13 978-0-00-753785-3

The authors assert their moral rights to be identified as the authors of this work.

British Library Cataloguing in Publication Data
A Catalogue record for this publication is available from the British Library.

Written by Rob Ellis, Kevin Evans, Keith Gordon, Chris Pearce, Trevor Senior, Brian Speed and Sandra Wharton
Commissioned by Katie Sergeant
Project managed by Elektra Media Ltd
Development edited and copy-edited by Gudrun Kaiser
Edited by Helen Marsden
Proofread by Grace Glendinning
Illustrations by Jerry Fowler and Ann Paganuzzi
Typeset by Elektra Media Ltd
Cover design by Angela English
Production by Emma Roberts

Printed and bound by RR Donnelley at Glasgow, UK

Acknowledgements
The publishers wish to thank the following for permission to reproduce photographs. Every effort has been made to trace copyright holders and to obtain their permission for the use of copyright materials. The publishers will gladly receive any information enabling them to rectify any error or omission at the first opportunity.

Cover gyn9037/Shutterstock.

Contents

8 Simplifying numbers

9 Interpreting data

10 Algebra

11 Congruence and scaling

12 Fractions and decimals

13 Proportion

14 Circles

15 Equations and formulae

16 Comparing data

Introduction

Maths Frameworking has been fully revised and updated in line with the 2014 Programme of study for Key Stage 3 Mathematics. This third edition provides complete coverage of the subject content of the new curriculum, as well as ample material to support pupils in fulfilling its three overarching aims: developing mathematical fluency, mathematical reasoning, and problem-solving skills.

Maths Frameworking 3rd edition components

- **Pupil Books:** three per year, catering for different levels of ability.
- **Teacher Packs:** one to accompany each Pupil Book.
- **Homework Books:** one per year, encompassing practice material, which is graduated in difficulty and suitable for all ability levels.
- **Intervention Workbooks:** five workbooks of increasing difficulty level (levels 3–7 of the previous national curriculum), designed for targeted one-to-one intervention to help pupils achieve the expected level of progress.
- **Digital resources on Collins Connect:** Collins Connect is an online platform which includes interactive versions of the Pupil Books, with interactive classroom and homework activities, assessments, tools and videos that have been specially developed to help pupils improve their mathematics skills.

Features of this Teacher Pack

This Teacher Pack accompanies *Maths Frameworking 3rd edition Pupil Book 2.2* and caters for middle-tier pupils (working at roughly Levels 5–7 of the previous curriculum). Lower-tier pupils (working at roughly Levels 4–6) are catered for by Pupil Book 2.1 and the accompanying Teacher Pack 2.1, and higher-tier pupils (Levels 5–7, with a greater proportion of level 6 and 7 materials) are catered for by Pupil Book 2.3 and Teacher Pack 2.3. The topics and their sequence are the same across all three books but are handled at different levels, allowing for sensitive differentiation in mixed-ability classes.

We have produced around 100 one-hour lessons to cover all the material you need in order to deliver the new Programme of study. These lessons should provide the flexibility to include tests, extended activities and revision classes in the teaching programme, and allow for the normal events that may disrupt teaching time. For further help when mapping lessons and producing your scheme of work, please consult the Contents pages and the Programme of study matching chart provided in this book.

Chapter overview

Pupil Book chapters are categorised according to the sections of the Programme of study to which they relate, and are colour-coded as follows:

- Number – pink
- Algebra – green
- Ratio, proportion and rates of change – lilac
- Geometry and measures – orange
- Probability – navy
- Statistics – teal

Each chapter of this Teacher Pack starts with an outline of the content covered in the chapter, explaining clearly how the lesson plans cover the National Curriculum and providing context for the topics covered. Its features are listed below.

- **Learning objectives** identify the key learning outcomes.
- **Prior knowledge** highlights the underpinning mathematics that pupils will draw on in the chapter, taking into account the coverage of the new Programme of study for Key Stage 2.
- **Context** provides some real-life or historical background for the key mathematical ideas explored in the chapter and gives suggestions on how to use the corresponding chapter opener in the Pupil Book.
- **Discussion points** provide some ideas for warm-up questions to introduce pupils to the topic.
- **Associated Collins ICT resources** provide an overview of the digital resources related to the chapter, which are available on the Collins Connect online platform.
- **Curriculum references** show how the materials meet the requirements of the new curriculum with references to the 2014 Programme of study.
- **Fast-track for classes following a 2-year scheme of work** provides suggestions for parts of the chapter that can be omitted if the class is following a condensed Key Stage 3.

Lesson plans

A lesson plan is provided for each topic in the Pupil Book. Every plan follows the same format, making it easy to prepare for and use. Its features are listed below.

- **Learning objectives** identify the key learning outcomes of the lesson.
- **Links to other subjects** highlight the topics' cross-curricular links, encouraging pupils to relate what they are learning in mathematics to what they are learning in other subjects.
- **Resources** list the content relating to the lesson that is available in other components of the scheme.
- **Homework activities** provide references to the related activities in the Homework Book and on the Collins Connect online platform.
- **Key words** provide a list of the key terms that pupils need to understand and use to talk about this topic. The words match the key words in the Pupil Book.
- **Problem solving, reasoning and financial skills help** details how the topic relates to the overarching aims of the Programme of study and, where appropriate, how it helps pupils to develop their financial skills.
- **Common misconceptions and remediation** pinpoints typical problems that pupils may have in getting to grips with the topic, and gives advice on how pupils can overcome these problems.
- **Probing questions** are designed to bring out important aspects of the topic and to link these aspects to pupils' prior knowledge.
- **Part 1** of the lesson plan is an engaging oral and mental starter involving the whole class; it has been designed to require minimal specialised equipment.
- **Part 2** is a main lesson activity, which helps you to lead pupils into the exercise questions.
- **Part 3** is a plenary designed to round off the three-part lesson.
- **Answers** are provided at the end of each lesson plan. Answers for the review questions and the activity are at the end of each chapter.

Answers for other course components

Answers to the Homework Book and Intervention Workbook questions can be found at this website link: **www.collins.co.uk/mathsframeworkinganswers**.

Scheme of work

A flexible 2-year and 3-year scheme of work is provided at the back of the book and in editable Word and Excel format on the CD-ROM accompanying this book. It shows two routes through the content, which are suitable for schools following a 2- or 3-year programme for Key Stage 3.

Teacher Pack CD-ROM

The CD-ROM in the front cover of this book contains:

- all the lesson plans in Word format for you to customise as you wish
- the 2- and 3-year schemes of work in Word and Excel format
- a printable version of the progression checklists for each chapter of the Pupil Book, which may be copied and distributed to the class for self-assessment, and for pupils to gauge their progress and learn how to improve
- skills maps for fluency, mathematical reasoning and problem solving
- a letter describing the philosophy and features of *Maths Frameworking 3rd edition* that can be customised and distributed to parents.

Digital resources available on Collins Connect

A variety of engaging resources are available on Collins Connect to support and enhance your teaching.

- **Classroom presentation resources** include:
    - an interactive version of the three Pupil Books for each year
    - interactive activities, including matching pairs and drag-and-drops, that can be used as starters and plenaries to improve fluency or stimulate group discussion
    - engaging video clips that develop pupils' conceptual understanding of key topics
    - 'Maths Man' videos that use rhythm and music to help pupils remember essential maths facts
    - exciting real-life videos and images, with accompanying teacher notes, that bring to life the awe and wonder of mathematics and encourage open conversations involving mathematics
    - audio-worked examples that model how to approach a variety of question types – including problem solving – and demonstrate best practice.
- Innovative **skills-building interactives** that enable pupils to explore and discover confidentially in order to build their conceptual understanding of, and skills in, proportional reasoning and equations.
- A **digital version of the Pupil Book glossary**, complete with recordings of all the key words – ideal for supporting literacy.
- **Automarked homework** tasks to accompany each lesson. Differentiated homeworks are provided for in the different Pupil Books.
- **Automarked assessments with diagnostic feedback**, including a diagnostic assessment for the start of the course, half-term assessments covering the main

chapter content, and an end-of-year test, which is split into two halves so that it fits easily into lesson time.

Collins Connect also provides invaluable tools to aid in tracking pupils' progress:

- **Synchronisation with SIMS**, enabling you to set up individual pupil accounts quickly and easily.
- **Task assignment** functionality, which lets you set any part of the interactive Pupil Book to a pupil or class as an assignment. As with the interactive homework tasks, you can set pupils a deadline for completion.
- **Results overviews** show the marks achieved by your classes in the homework tasks and tests you have set for them. Results can be displayed by a pupil, an assignment or a class, so that you can pinpoint areas for intervention easily.

Find out more at **www.collins.co.uk/connect**.

Maths Frameworking and the 2014 Key Stage 3 Programme of study for mathematics

Rationale

Following the publication of the Schools White Paper, 'The Importance of Teaching' (DfE, November 2010), the UK government instigated a review into the national curriculum in mathematics. The stated aim of the review was to benchmark expectations for pupils in this country against the expectations of the most successful nations in the world, and to draw on this information to design a new curriculum that would put English pupils on a par with pupils in the highest performing countries. In February 2013, the DfE began an extensive consultation process involving the mathematics community. In September 2013, this process resulted in the publication of a revised Programme of study for Key Stage 3 mathematics. The new curriculum became statutory in September 2014.

The 2014 Programme of study for mathematics

Subject content

The subject content of both the new Key Stage 2 and Key Stage 3 programmes of study has changed with respect to the former curriculum. Probability has been removed from the Key Stage 2 curriculum, in favour of greater emphasis on the foundations of algebra and on skills in number (for example, multiplication tables up to 12×12 and long multiplication and division). This means that the expectations of the prior knowledge that pupils bring with them at the start of Key Stage 3 will also change. Although there has been some reduction of content, in other areas the expectation in Key Stage 2 is very demanding as it includes some topics that were at level 6 in the old Programme of study. This third edition of *Maths Frameworking* considers these new expectations and creates a framework for progression from this new starting point.

At Key Stage 3, the Programme of study is divided into six strands. Within these, some of the content requirements have been more fully described. One example is a deeper exploration of the properties of prime numbers than before. There is also increased precision in the areas of algebra and of geometry and measures. 'Ratio, proportion and rates of change' now stands alone as a strand in its own right, emphasising the links between fractions, ratios, proportion and percentages. Probability and Statistics are now also treated as separate strands. Some of the material that was at level 8 in the old Programme of study has been removed. Trigonometry is an example. The intention is that pupils who grasp concepts rapidly should be challenged with more demanding problems on the same subject matter rather than being accelerated through new content in preparation for GCSE. The structure and coverage of the *Maths Frameworking* Pupil Books has been reworked comprehensively to reflect these changes.

Overarching aims

A key difference in the new Programme of study for mathematics is the introduction of three overarching aims, which are presented as integral to study of the subject content. The aims are to ensure that all pupils:

- become fluent in the fundamentals of mathematics, including through varied and frequent practice with increasingly complex problems over time, so that pupils develop conceptual understanding and the ability to recall and apply knowledge rapidly and accurately.
- reason mathematically by following a line of enquiry, conjecturing relationships and generalisations, and developing an argument, justification or proof using mathematical language.
- can solve problems by applying their mathematics to a variety of routine and non-routine problems with increasing sophistication, including breaking down problems into a series of simpler steps and persevering in seeking solutions.

The Key Stage 2 and Key Stage 3 programmes of study share the same overarching aims, and these aims form the basis for the assessment objectives in the new mathematics GCSE for first teaching in September 2015. The intention of the DfE is to ensure smooth progression through the three Key Stages, with the Key Stage 3 Programme of study consolidating and building on understanding from Key Stage 2 in order to form solid foundations for further study at GCSE and beyond.

The aims of the national curriculum are intended to be developed and applied across the mathematical content of the Programme of study, and their treatment in *Maths Frameworking 3rd edition* reflects this. For example, pupils develop fluency by using algebra to consolidate their understanding from Key Stage 2 and generalise the structure of arithmetic. They reason mathematically and extend their understanding of the number system by making connections between number relationships and their algebraic and graphical representations. They solve problems by selecting appropriate concepts, methods and techniques to apply to unfamiliar and non-routine problems.

Fluency and linking

Fluency in mathematics has two important facets: familiarity and recollection. Familiarity is important because it means that, when faced with a calculation, pupils can recognise the steps required to complete it by drawing on prior experience of completing other similar calculations. The ability to recollect pertinent mathematics facts and techniques further enhances pupils' ability to complete calculations rapidly, as they access, for example, their knowledge of times tables and mental mathematics techniques to reach an answer without needing to use a calculator.

Fundamental to fostering both of these aspects of mathematical fluency, is ensuring that pupils gain extensive practice. Rich and varied practice lies at the heart of *Maths*

Frameworking. At the start of each lesson plan in this Teacher Pack, we suggest a series of probing questions, which are designed to encourage pupils to recall previously learnt mathematics facts and skills that are of relevance to the topic. The exercises within every topic offer a wide range of questions of increasing complexity to which pupils can apply these skills and knowledge. The Homework Books and auto-marked homework activities on Collins Connect offer further opportunities for practice.

In addition to ensuring plentiful practice, the *Maths Frameworking* scheme ensures that the practice encourages in pupils the understanding and mastery that are the hallmarks of true fluency by providing opportunities for practice across different mathematical and cross-curricular contexts.

In Ofsted's paper, 'Mathematics: Understanding the Score' (Ofsted, September 2008), Ofsted identified that the majority of pupils had too few opportunities to use and apply mathematics and to make connections across different areas of the subject. In the 2014 national curriculum, mathematical topics figure in a wide variety of curricular subjects. For example, Design and technology pupils need to be comfortable when working with 2D and 3D plans, while data-handling skills are required in Geography, Science and Computing. The 2014 national curriculum also places renewed emphasis on financial education, with a focus within the new mathematics Programme of study on solving problems involving percentage increases and decreases, simple interest and repeated growth. This has clear links to the Programme of study for Citizenship, which requires pupils to learn the functions and uses of money, the importance and practice of budgeting, and managing risk.

Maths Frameworking draws out these intra- and cross-curricular links in order to help pupils to become adept at making connections. Throughout the books, questions are presented in varied contexts taken both from real life and from the subject areas mentioned previously, plus other subject areas. In the 'Review' questions at the end of each chapter, as well as checking pupils' grasp of the mathematics topics covered within it, other parts of mathematics are also brought into play. Questions are posed, which require pupils to combine learning across different topics in order to answer them. Furthermore, each chapter begins with an opening page that presents real-life applications of the topic covered, giving pupils a window onto why it is important and useful outside of the classroom. This helps pupils to make links between the mathematics they are studying in different parts of the curriculum, to the relevance in real life, rather than learning only the mathematics. These Pupil Book features are supported by accompanying guidance in this Teacher Pack: the 'Context' section at the start of each chapter provides background information for teachers, while the 'Links to other subjects' section of each lesson plan highlights cross-curricular links in each topic.

Problem solving and reasoning

The ability to solve problems within mathematics and in other subjects is a key skill, which consultations have shown that employers require across different industries and economic sectors. In the specifications for the new mathematics GCSE for first teaching in September 2015, problem solving is to become one of the three assessment objectives, together with using and applying standard techniques and reasoning mathematically. Teaching at Key Stage 3 needs to build pupils' problem-solving skills in preparation for study at GCSE level. Crucial to this is equipping pupils with strategies that will help them to see the 'big picture' and nurture mathematical independence, allowing time for thinking, and encouraging discussion in the classroom. The lesson plans in this book provide a wide variety of suggestions on how teachers can do this, and offer varied opportunities for probing pupils' understanding and encouraging reflection.

The first step in problem solving in mathematics at this level is often identifying 'where the maths is' and what mathematical techniques can be applied to answer a question. The problem-solving questions in *Maths Frameworking*, especially the extended double-page problem-solving spreads, guide pupils through this process, equipping them with techniques to decipher word problems and 'translate' descriptions into mathematics. Making conjectures and trying different approaches to solve problems is also an important reflex that *Maths Frameworking* develops by providing guided questions in which pupils can experiment and are encouraged to question and reflect. Across the varied problem-solving questions in *Maths Frameworking*, pupils are also encouraged to develop visual representation techniques by using graphs, charts or diagrams to help them get to grips with a word problem or identify a pattern.

Mathematical reasoning is another key skill that has been embedded in this third edition of *Maths Frameworking*. Pupils are reasoning mathematically when they follow a line of enquiry, conjecture a relationship or generalisation, and develop an argument, justification or proof using mathematical language. Reasoning questions and extended double-page reasoning spreads, give pupils opportunities to do this, encouraging them to develop the skill set needed in order to investigate, prove and explain. Developing good habits for showing working and justifying answers will stand pupils in good stead for the second new assessment objective at GCSE level, for which they will need to demonstrate their ability to: make deductions, inferences and draw conclusions from mathematical information, construct chains of reasoning to achieve a given result, interpret and communicate information accurately, present arguments and proofs and assess the validity of an argument and critically evaluate a given way of presenting information. Giving pupils the opportunity to improve the quality of their reasoning is also important in developing their abilities in this area. The 'Common misconceptions and remediation' section of each lesson plan aims to guide teachers in helping pupils to improve.
Providing opportunities for extended responses builds mathematical resilience in pupils, and is a necessary skill for problem solving and reasoning. In the Pupil Book, particular types of question are flagged so that teachers and pupils can find them easily and see which skills are being worked on:

PS indicates questions that require pupils to solve problems

MR indicates questions that require pupils to reason mathematically

FS indicates questions that help to develop pupils' financial skills and financial awareness.

Approach to progression

In the 2014 curriculum, the system of levels formerly used to report pupils' attainment and progress has been removed. The reason given for this by the DfE was that the over-emphasis on levelling in schools encouraged teachers to focus too tightly on 'hitting the numbers' rather than forming a more holistic view of each pupil's progress in terms of conceptual understanding. Ofsted subject-specific exemplification states that the 'development of all pupils' conceptual understanding [...] and progression within each lesson and over time' is one of the hallmarks of outstanding mathematical teaching. In keeping with this, the new Programme of study sets out what subject knowledge should be acquired by the end of Key Stage 3, but gives schools the freedom to develop their own scheme of work for covering it, and their own system for tracking pupils' progress.

The scheme of work and lesson plans in *Maths Frameworking* offer schools a framework structured to ensure progression both over time and within every lesson. The scheme of work has been developed to ensure that all pupils meet expectations in relation to the aims and content of the Key Stage 3 Programme of study, and that they have the opportunity to progress through the curriculum in a way that supports understanding and challenge. We appreciate that giving pupils the *appropriate degree* of challenge for their current level of ability and knowledge is crucial to their making progress, and the three different tiers of the *Maths Frameworking* scheme enable us to achieve this by creating differentiated progression pathways for pupils according to their different starting points. Pupil Book 2.1 covers levels 4–6 of the former national curriculum; Pupil Book 2.2 covers levels 5–7, and Pupil Book 2.3 covers levels 5–7, with a greater proportion of level 6 and 7 materials. We have consulted with mathematics teachers and experts on how material should be presented and structured within the different tiers in order to ensure smooth progression for different pupils. The result is materials that are tailored to pupils' differential prior knowledge, skills and understanding, which gradually build on these to foster and support progress.

Progression is built into every exercise in the scheme. Exercises start with straightforward questions to consolidate skills and understanding; then move on to more varied and demanding material, and end with extension tasks that are designed to challenge and stretch pupils at the level that is appropriate to their learning abilities. At the end of each chapter of the Pupil Book, pupils are encouraged to reflect on what they have learned and what they need to do to progress by self-assessing using the 'Ready to progress?' checklist (also available in printable format on the CD-ROM accompanying this book).

To enable schools and teachers to check that pupils are on track to meet expectations, we have developed a system of colour-coded icons to show the level of difficulty of each question in the Pupil Books. Teachers and pupils can see at a glance whether they are working in line with expected progress for each learning pathway and year of study.
We appreciate that, during the first stages of implementing the new curriculum, many schools will continue to monitor pupils' progression in terms of the levels used in the former curriculum. To facilitate this, and aid in the transition between this system and the new system, which is based on expectations of progress, we have based our system of icons on the former level descriptors. The table below shows the approximate correlation between the progress indicators used in *Maths Frameworking* and the levels used in the previous national curriculum:

	Less than expected progress	Expected progress	More than expected progress
Pupil Book 1.1	≤ Level 3	Level 4	≥ Level 5
Pupil Book 1.2	≤ Level 4	Level 5	≥ Level 6
Pupil Book 1.3	≤ Level 4	Level 5	≥ Level 6
Pupil Book 2.1	≤ Level 4	Level 5	≥ Level 6
Pupil Book 2.2	≤ Level 5	Level 6	≥ Level 7
Pupil Book 2.3	≤ Level 5	Level 6	≥ Level 7
Pupil Book 3.1	≤ Level 4	Level 5	≥ Level 6
Pupil Book 3.2	≤ Level 5	Level 6	≥ Level 7
Pupil Book 3.3	≤ Level 6	Level 7	≥ Level 8

Another aspect of progression that has been emphasised in recent studies on mathematics pedagogy is the CPA, or concrete-pictorial-abstract, sequence. Research suggests that progression in mathematical thinking rests on transitioning from the physical manipulation of concrete materials (an important feature of practice at primary level) to pictorial representations, and finally to written representations in mathematical notation. Developing connections between these three types of experiences, and thereby transitioning through the three stages of understanding, is foremost for making solid progress and becoming mathematically confident and fluent. *Maths Frameworking* aids pupils in building on their experiences in Key Stage 2 and progressing to pictorial and abstract representations by drawing on a variety of different media, for example, pictures, diagrams and charts, as well as correct mathematical notation, in order to explain concepts. Where appropriate, for concepts that are being introduced for the first time, the lesson plans in this Teacher Pack provide suggestions for guidance to use in class.

Approach to assessment

In its paper, 'Mathematics: Understanding the Score', Ofsted identified the need for teachers to place greater emphasis on developing pupils' understanding and on checking it throughout lessons. Ofsted advocated that schools should ensure that pupils have a wide range of opportunities to use and apply mathematics, and that these should be underpinned by thorough assessment, recording and reporting. Regular assessment is known to be a motivating factor for pupils, as gaining awareness of the progress they are making helps to build their confidence, thereby laying the foundations for further achievement. The abolition of levels in the 2014 national curriculum will see a phase shift in how schools approach assessment; *Maths Frameworking* offers a range of tools to support schools in making this transition.

Under the new curriculum, it is the intention of the DfE that assessment should be built into and become integral to the school curriculum, thus allowing schools to check what pupils have learned and whether they are on track to meet expectations at the end of the Key Stage ('Assessing Without Levels', DfE, June 2013). This means creating opportunities for formal, periodic assessment and building informal assessment into daily classroom practices through, for example, the effective use of probing questions to assess progress and identify and tackle misconceptions.

Maths Frameworking supports schools' own approaches to formative assessment by providing a suite of resources that will enable them to track and offer evidence of each pupil's progress through the Key Stage, however and whenever they desire.

- Discussion points are suggested at the start of each chapter of this Teacher Pack, to enable teachers to detect prior knowledge before starting work on a topic.
- Probing questions are provided in each lesson plan to support informal assessment during the course of normal lessons.
- In order to facilitate regular monitoring of pupils' progress, at the end of each chapter of the Pupil Book, synoptic review questions are provided so that pupils and teachers have an opportunity to assess understanding of the topic covered.
- The 'Ready to progress' chart at the end of each chapter provides an opportunity for self-assessment, and for pupils to see what they need to do next in order to improve.
- Auto-marked homework tasks on the Collins Connect platform provide opportunities for formative assessment for each topic in the Pupil Books. Diagnostic feedback enables pupils to identify areas that they need to strengthen and directs them to resources that will help them. Results overviews help you to identify pupils' strengths and any areas where you might need to intervene.
- Tailor-made assessments are also provided for each tier of the *Maths Frameworking* scheme on the Collins Connect platform. A diagnostic assessment is available to evaluate pupils' starting points at the beginning of the first year of Key Stage 3 and to help you to decide how to set them. Thereafter, differentiated half-term tests are provided for each strand of the scheme, each one covering roughly five to six chapters of material. These tests evaluate pupils' progress against the stated objectives for each topic and again provide diagnostic feedback to identify pupils' strengths and areas for improvement.

Maths Frameworking and the new mathematics GCSE

In the new national curriculum, there is greater continuity between the Key Stage 3 Programme of study and the subject content of the new maths GCSE for first teaching in September 2015. The same six thematic strands are used in both Key Stages 3 and 4, and there are a number of examples where the wording used is exactly the same in the subject content for both levels. This means that pupils who have a secure understanding of the material in the Key Stage 3 Programme of study will be very well prepared for the study of mathematics at Key Stage 4.

The assessment objectives for the new mathematics GCSE also build on and formalise the overarching aims of the Programme of study for Key Stages 1, 2 and 3, thereby forming a unified pathway for mathematics across primary and secondary. The Key Stage 3 Programme of study makes it clear, however, that there is no expectation that teaching of content from previous Key Stages should be repeated. Instead, pupils should be given the opportunity in Key Stage 4 to make choices and decisions about the mathematics they use, drawing on a well-developed toolkit to interpret and communicate mathematics for different audiences and purposes. It is this well-developed toolkit – a secure grounding in both content and process skills – that *Maths Frameworking 3rd edition* provides, thus creating the firm foundations that should enable every pupil to progress to GCSE and succeed at it.

Programme of study matching chart

This chart matches the subject content of the 2014 Programme of study for Key Stage 3 mathematics to specific lesson plans contained in this Teacher Pack.

Working mathematically Through the mathematics content, pupils should be taught to:	**Chapter title**	**Lesson number and title**
Number		
Understand and use place value for decimals, measures and integers of any size	Covered in Year 7	
Order positive and negative integers, decimals and fractions; use the number line as a model for ordering of the real numbers; use the symbols =, ≠, <, >, ≤, ≥	12 Fractions and decimals	12.1 Adding and subtracting fractions 12.2 Multiplying fractions and integers 12.3 Dividing with integers and fractions 12.4 Multiplication with large and small numbers 12.5 Division with large and small numbers
Use the concepts and vocabulary of prime numbers, factors (or divisors), multiples, common factors, common multiples, highest common factor, lowest common multiple, prime factorisation, including using product notation and the unique factorisation property	1 Working with numbers	1.2 Factors and highest common factors (HCF) 1.3 Lowest common multiples (LCM) 1.5 Prime factors
Use the four operations, including formal written methods, applied to integers, decimals, proper and improper fractions, and mixed numbers, all both positive and negative	1 Working with numbers 12 Fractions and decimals	1.1 Multiplying and dividing negative numbers 1.2 Factors and highest common factors (HCF) 1.3 Lowest common multiples (LCM) 1.5 Prime factors 12.1 Adding and subtracting fractions 12.2 Multiplying fractions and integers 12.3 Dividing with integers and fractions 12.4 Multiplication with large and small numbers 12.5 Division with large and small numbers
Use conventional notation for the priority of operations, including brackets, powers, roots and reciprocals	Covered in Year 7	

Recognise and use relationships between operations including inverse operations	15 Equations and formulae	15.1 Equations with brackets 15.2 Equations with the variable on both sides 15.3 More complex equations 15.4 Rearranging formulae
Use integer powers and associated real roots (square, cube and higher), recognise powers of 2, 3, 4, 5 and distinguish between exact representations of roots and their decimal approximations	1 Working with numbers	1.4 Powers and roots
Interpret and compare numbers in standard form $A \times 10^n$ $1 \le A < 10$, where *n* is a positive or negative integer or zero	8 Simplifying numbers	8.1 Powers of 10 8.2 Large numbers and rounding 8.4 Standard form with large numbers 8.5 Multiplying with numbers in standard form
Work interchangeably with terminating decimals and their corresponding fractions (such as 3.5 and $\frac{7}{2}$)	Covered in Year 9	
Define percentage as 'number of parts per hundred', interpret percentages and percentage changes as a fraction or a decimal, interpret these multiplicatively, express one quantity as a percentage of another, compare two quantities using percentages, and work with percentages greater than 100%	4 Percentages	4.1 Calculating percentages 4.2 Calculating percentage increases and decreases 4.3 Calculating a change as a percentage
Interpret fractions and percentages as operators	4 Percentages	4.1 Calculating percentages 4.2 Calculating percentage increases and decreases 4.3 Calculating a change as a percentage
Use standard units of mass, length, time, money and other measures, including with decimal quantities	Covered in Year 7	
Round numbers and measures to an appropriate degree of accuracy [for example, to a number of decimal places or significant figures]	8 Simplifying numbers	8.2 Large numbers and rounding 8.3 Significant figures

Use approximation through rounding to estimate answers and calculate possible resulting errors expressed using inequality notation $a < x \leq b$	Covered in Year 7	
Use a calculator and other technologies to calculate results accurately and then interpret them appropriately	Covered in Year 7	
Appreciate the infinite nature of the sets of integers, real and rational numbers	5 Sequences	5.1 Using flow diagrams to generate sequences 5.2 The nth term of a sequence 5.3 Working out the nth term of a sequence
Algebra		
Use and interpret algebraic notation, including: *ab* in place of $a \times b$ 3*y* in place of $y + y + y$ and $3 \times y$ a^2 in place of $a \times a$, a^3 in place of $a \times a \times a$; a^2b in place of $a \times a \times b$ $\frac{a}{b}$ in place of $a \div b$ coefficients written as fractions rather than as decimals brackets	10 Algebra	10.1 Algebraic notation 10.2 Like terms 10.3 Expanding brackets 10.4 Using algebraic expressions 10.5 Using index notation
Substitute numerical values into formulae and expressions, including scientific formulae	7 Graphs 10 Algebra	7.1 Graphs from linear equations 7.4 Real-life graphs 10.4 Using algebraic expressions
Understand and use the concepts and vocabulary of expressions, equations, inequalities, terms and factors	10 Algebra	10.1 Algebraic notation 10.2 Like terms 10.3 Expanding brackets 10.4 Using algebraic expressions
Simplify and manipulate algebraic expressions to maintain equivalence by: collecting like terms multiplying a single term over a bracket taking out common factors expanding products of two or more binomials	10 Algebra	10.2 Like terms 10.3 Expanding brackets 10.4 Using algebraic expressions
Understand and use standard mathematical formulae; rearrange formulae to change the subject	15 Equations and formulae	15.3 More complex equations 15.4 Rearranging formulae

Model situations or procedures by translating them into algebraic expressions or formulae and by using graphs	7 Graphs	7.4 Real-life graphs
Use algebraic methods to solve linear equations in one variable (including all forms that require rearrangement)	15 Equations and formulae	15.1 Equations with brackets 15.2 Equations with the variable on both sides 15.3 More complex equations 15.4 Rearranging formulae
Work with coordinates in all four quadrants	7 Graphs	7.1 Graphs from linear equations 7.2 Gradient (steepness) of a straight line 7.3 Graphs from simple quadratic equations 7.4 Real-life graphs
Recognise, sketch and produce graphs of linear and quadratic functions of one variable with appropriate scaling, using equations in *x* and *y* and the Cartesian plane	7 Graphs	7.1 Graphs from linear equations 7.2 Gradient (steepness) of a straight line 7.3 Graphs from simple quadratic equations 7.4 Real-life graphs
Interpret mathematical relationships both algebraically and graphically		
Reduce a given linear equation in two variables to the standard form $y = mx + c$; calculate and interpret gradients and intercepts of graphs of such linear equations numerically, graphically and algebraically	7 Graphs	7.1 Graphs from linear equations 7.2 Gradient (steepness) of a straight lines
Use linear and quadratic graphs to estimate values of *y* for given values of *x* and vice versa and to find approximate solutions of simultaneous linear equations	Covered in Year 7	
Find approximate solutions to contextual problems from given graphs of a variety of functions, including piece-wise linear, exponential and reciprocal graphs	7 Graphs	7.4 Real-life graphs
Generate terms of a sequence from either a term-to-term or a position-to-term rule	5 Sequences	5.1 Using flow diagrams to generate sequences 5.2 The *n*th term of a sequence 5.3 Working out the *n*th term of a sequence 5.4 The Fibonacci sequence

Recognise arithmetic sequences and find the *n*th term	5 Sequences	5.2 The *n*th term of a sequence 5.3 Working out the *n*th term of a sequence
Recognise geometric sequences and appreciate other sequences that arise	5 Sequences	5.1 Using flow diagrams to generate sequences 5.4 The Fibonacci sequence
Ratio, proportion and rates of change		
Change freely between related standard units [for example time, length, area, volume/capacity, mass]	Covered in Year 7	
Use scale factors, scale diagrams and maps	11 Congruence and scaling	11.2 Enlargements 11.3 Shape and ratio 11.4 Scales
Express one quantity as a fraction of another, where the fraction is less than 1 and greater than 1	Covered in Year 7	
Use ratio notation, including reduction to simplest form	11 Congruence and scaling	11.3 Shape and ratio 11.4 Scales
Divide a given quantity into two parts in a given part : part or part : whole ratio; express the division of a quantity into two parts as a ratio	Covered in Year 7	
Understand that a multiplicative relationship between two quantities can be expressed as a ratio or a fraction	11 Congruence and scaling	11.3 Shape and ratio
Relate the language of ratios and the associated calculations to the arithmetic of fractions and to linear functions	Covered in Year 7	
Solve problems involving percentage change, including: percentage increase, decrease and original value problems and simple interest in financial mathematics	Covered in Year 7	
Solve problems involving direct and inverse proportion, including graphical and algebraic representations	13 Proportion	13.1 Direct proportion 13.2 Graphs and direct proportion 13.3 Inverse proportion 13.4 Comparing direct proportion and inverse proportion
Use compound units such as speed, unit pricing and density to solve problems	Covered in Year 9	

Geometry and measures		
Derive and apply formulae to calculate and solve problems involving: perimeter and area of triangles, parallelograms, trapezia, volume of cuboids (including cubes) and other prisms (including cylinders)	6 Area of 2D and 3D shapes	6.1 Area of a triangle 6.2 Area of a parallelogram 6.3 Area of a trapezium 6.4 Surface areas of cubes and cuboids
Calculate and solve problems involving: perimeters of 2D shapes (including circles), areas of circles and composite shapes	6 Area of 2D and 3D shapes 14 Circles	6.1 Area of a triangle 6.2 Area of a parallelogram 6.3 Area of a trapezium 14.1 The circle and its parts 14.2 Circumference of a circle 14.3 Formula for the circumference of a circle 14.4 Formula for the area of a circle
Draw and measure line segments and angles in geometric figures, including interpreting scale drawings	Covered in Year 7	
Derive and use the standard ruler and compass constructions (perpendicular bisector of a line segment, constructing a perpendicular to a given line from/at a given point, bisecting a given angle); recognise and use the perpendicular distance from a point to a line as the shortest distance to the line	2 Geometry	2.5 Constructions
Describe, sketch and draw using conventional terms and notations: points, lines, parallel lines, perpendicular lines, right angles, regular polygons, and other polygons that are reflectively and rotationally symmetric	2 Geometry	2.1 Angles in parallel lines 2.2 The geometric properties of quadrilaterals 2.5 Constructions
Use the standard conventions for labelling the sides and angles of triangle ABC, and know and use the criteria for congruence of triangles	11 Congruence and scaling	11.1 Congruent shapes
Derive and illustrate properties of triangles, quadrilaterals, circles, and other plane figures [for example, equal lengths and angles] using appropriate language and technologies	2 Geometry 14 Circles	2.2 The geometric properties of quadrilaterals 14.1 The circle and its parts 14.2 Circumference of a circle 14.3 Formula for the circumference of a circle 14.4 Formula for the area of a circle

Identify properties of, and describe the results of, translations, rotations and reflections applied to given figures	2 Geometry 11 Congruence and scaling	2.3 Rotations 2.4 Translations 11.1 Congruent shapes
Identify and construct congruent triangles, and construct similar shapes by enlargement, with and without coordinate grids	11 Congruence and scaling	11.1 Congruent shapes 11.2 Enlargements
Apply the properties of angles at a point, angles at a point on a straight line, vertically opposite angles	2 Geometry	2.1 Angles in parallel lines
Understand and use the relationship between parallel lines and alternate and corresponding angles	2 Geometry	2.1 Angles in parallel lines
Derive and use the sum of angles in a triangle and use it to deduce the angle sum in any polygon, and to derive properties of regular polygons	Covered in Year 9	
Apply angle facts, triangle congruence, similarity and properties of quadrilaterals to derive results about angles and sides, including Pythagoras' Theorem, and use known results to obtain simple proofs	Covered in Year 7	
Use Pythagoras' Theorem and trigonometric ratios in similar triangles to solve problems involving right-angled triangles	Covered in year 9	
Use the properties of faces, surfaces, edges and vertices of cubes, cuboids, prisms, cylinders, pyramids, cones and spheres to solve problems in 3D	6 Area of 2D and 3D shapes	6.4 Surface areas of cubes and cuboids
Interpret mathematical relationships both algebraically and geometrically		
Probability		
Record, describe and analyse the frequency of outcomes of simple probability experiments involving randomness, fairness, equally and unequally likely outcomes, using appropriate language and the 0–1 probability scale	3 Probability	3.1 Probability scales 3.4 Experimental probability
Understand that the probabilities of all possible outcomes sum to 1	3 Probability	3.1 Probability scales 3.2 Mutually exclusive outcomes 3.3 Using a sample space to calculate probabilities

Enumerate sets and unions/intersections of sets systematically, using tables, grids and Venn diagrams	3 Probability	3.2 Mutually exclusive outcomes 3.4 Experimental probability
Generate theoretical sample spaces for single and combined events with equally likely, mutually exclusive outcomes and use these to calculate theoretical probabilities	3 Probability	3.2 Mutually exclusive outcomes 3.3 Using a sample space to calculate probabilities
Statistics		
Describe, interpret and compare observed distributions of a single variable through: appropriate graphical representation involving discrete, continuous and grouped data; and appropriate measures of central tendency (mean, mode, median) and spread (range, consideration of outliers)	9 Interpreting data 16 Comparing data	9.4 Creating scatter graphs Challenge: Football attendances 16.1 Grouped frequency tables 16.2 Drawing frequency diagrams 16.3 Comparing data 16.4 Which average to use?
Construct and interpret appropriate tables, charts, and diagrams, including frequency tables, bar charts, pie charts, and pictograms for categorical data, and vertical line (or bar) charts for ungrouped and grouped numerical data	9 Interpreting data 16 Comparing data	9.1 Pie charts 9.2 Creating pie charts 16.1 Grouped frequency tables 16.2 Drawing frequency diagrams
Describe simple mathematical relationships between two variables (bivariate data) in observational and experimental contexts and illustrate using scatter graphs	9 Interpreting data	9.3 Scatter graphs and correlation 9.4 Creating scatter graphs

1 Working with numbers

Learning objectives

- How to multiply and divide negative numbers
- How to find the highest common factor and the lowest common multiple of sets of numbers
- How to use powers and find roots
- How to find the prime factors of a number

Prior knowledge

- How to add and subtract negative integers
- How to order the operations following BIDMAS
- What a factor is
- What a multiple is

Context

- In this chapter, Lesson 1.1 recalls addition and subtraction of negative numbers and moves on to multiplying and dividing negative numbers. Lessons 1.2 and 1.3 introduce highest common factor (HCF) and lowest common multiple (LCM), respectively. The chapter moves on to powers and roots in Lesson 1.4. Lesson 1.5 introduces pupils to prime factors. The relationship of prime factors with HCF and LCM is explored briefly. Previous work on sequences is revisited and extended to include multiplicative sequences. The activity at the end of the chapter encourages thinking skills and the use of algebra to represent general solutions to a practical problem.

Discussion points

- Can all whole numbers be represented by a product of prime numbers?
- Why, when you multiply or divide two negative numbers, do you get a positive answer?
- Why, when you multiply or divide a negative number and a positive number, do you get a negative answer?

Associated Collins ICT resources

- Chapter 1 interactive activities on Collins Connect online platform

Curriculum references

Reason mathematically

- Extend their understanding of the number system; make connections between number relationships, and their algebraic and graphical representations

Solve problems

- Develop their mathematical knowledge, in part through solving problems and evaluating the outcomes, including multi-step problems

Number

- Use the concepts and vocabulary of prime numbers, factors (or divisors), multiples, common factors, common multiples, highest common factor, lowest common multiple, prime factorisation, including using product notation and the unique factorisation property
- Use the four operations, including formal written methods, applied to integers, decimals, proper and improper fractions, and mixed numbers, all both positive and negative
- Use integer powers and associated real roots (square, cube and higher), recognise powers of and distinguish between exact representations of roots and their decimal approximations

Fast-track for classes following a 2-year scheme of work

- Much of this material will be new to Year 8 pupils. Pupils can leave out questions 1 and 2 in Exercise 1A, which was covered in Year 7. If pupils are quick to grasp the concepts in this chapter they can move swiftly through the exercises, leaving out some of the questions.

Lesson 1.1 Multiplying and dividing negative numbers

Learning objective

- To carry out multiplications and divisions involving negative numbers

Resources and homework

- Pupil Book 2.2, pages 7–10
- Intervention Workbook 1, pages 23–25
- Intervention Workbook 3, pages 20–21
- Homework Book 2, section 1.1
- Online homework 1.1, questions 1–10

Links to other subjects

- **Science** – to understand magnetism and electricity

Key words

- negative number
- positive number

Problem solving and reasoning help

- The challenge question at the end of Exercise 1A draws on pupil's knowledge regarding simplifying like terms. It may be necessary to revisit this for **less able** pupils. **More able** pupils could design their own algebraic magic square.

Common misconceptions and remediation

- One of the main misconceptions pupils have when multiplying two negative numbers is giving a negative answer. Reinforce the fact that when multiplying two negative numbers, the answer will always be positive. And, when multiplying two numbers, pupils often think that the sign of the answer is determined by the sign of the largest number. Make sure that pupils do not rush through their work, as they need to have a clear understanding of the rules.

Probing questions

- Can you make up some multiplication/division questions to give an answer of –24?
- What keys could you have pressed on your calculator to get the answer of –144?
- Talk me through how you found the answers to the two questions.

Part 1

- Use a number line that you have drawn on the board, or a 'counting stick' with 10 divisions marked on it. The right end should show the number 0.
- Point out to pupils that as they look at the line, the values to the left of zero are negative.
- Give a value to each segment, say –3. As a class, count down the line in steps of –3 from zero. Point out the positions on the line as you say each number.
- Repeat with other values for the segments, for example: –4, –2, –1.5.
- Now give a value to each segment, say –6. Point at a position on the stick, for example, the fourth division, asking what value it represents.
- Repeat with other values for each segment and different positions on the stick.
- Explain that there is an easy way of finding the value at any position on the stick without counting down in steps. Move on to Part 2.

Part 2

- Draw a number line on the board and mark it from –10 to +10. Recall the rules for dealing with directed number problems using the number line. Make sure you recall that two signs together can be rewritten as one sign, for example: + + is +; + – is –; – + is –; – – is +. Another way to emphasise this is to say that if the signs are the same, then the overall sign is plus and if the signs are different, then it is minus.

- Demonstrate this by using the number line to work out: 7 + –3 (= +4) and –4 – –5 (= +1).
- Now ask for the answer to: –2 + –2 + –2 + –2 + –2 (= –10). Ask if there is another way to write this, that is: 5 × –2 (recall that multiplication is repeated addition).
- Repeat with other examples such as: – –4 – –4 – –4 = –3 × –4 (= +12).
- Ask pupils if they can see a quick way to work out products such as: –2 × + 3 or –5 × –4 or +7 × +3. Pupils should mention that the rule is the product of the numbers, combined with the rules they met earlier about combining signs.
- The – × – = + can cause problems. Ask pupils to complete this pattern: +2 × –3 = –6; +1 × –3 = –3; 0 × ... = ...; –1 × ... = ... , and so on. Link this to division, for example, if: –3 × +6 = –18, then –18 ÷ –3 = +6; if +5 × –3 = –15, then –15 ÷ +5 = –3.
- Ask pupils to explain a quick way to do these. As for multiplication, the numbers are divided; the sign of the final answer depends on the signs in the original division problem.
- **Pupils can now do Exercise 1A from Pupil Book 2.2.**

Part 3

- Ask some mental questions such as: How many negative fours make negative 16? What is: 6 – 9; –5 – +3; –4 – 3; –2 × +7; –32 ÷ –8; –3 squared?
- Encourage pupils to 'say the problem to themselves', for example, for +7 – –2, say 'plus seven minus minus two'.
- Make sure that pupils overcome the confusion about 'two negatives make a positive'. For example, pupils will often say: –6 – 7 = +13.

Answers

Exercise 1A

1 **a** 1 **b** –9 **c** 7 **d** 0 **e** –8 **f** –10 **g** –10 **h** 1 **i** –18 **j** –2

2

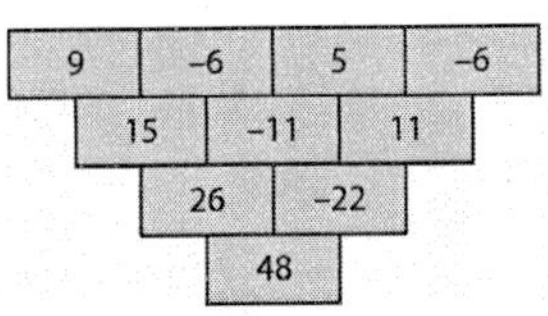

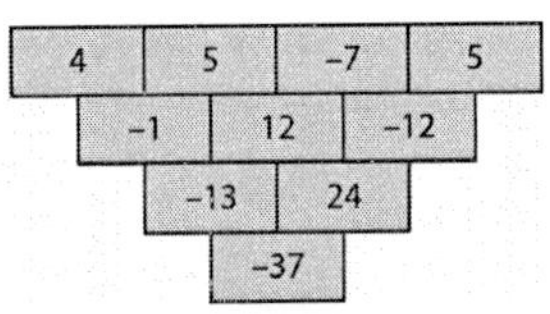

3 **a** –6 **b** –12 **c** –10 **d** 18
e –24 **f** –20 **g** 12 **h** 6
i –14 **j** –16 **k** 60 **l** –32
m 30 **n** –18 **o** 16 **p** 56

4 For example 3 × –4, –2 × 6, –1 × 12, 12 ÷ –1, –12 ÷ 1

5 **a** –4 **b** –6 **c** –3 **d** 2 **e** –4 **f** –8 **g** 8 **h** 6
i –3.5 **j** 2 **k** –6 **l** 2 **m** 7.5 **n** –9 **o** 4 **p** –4.5

6 **a**

6	–9	12	–15
–12	18	–24	30
4	–6	8	–10
–10	15	–20	25

b

×	–1	–3	4	–6
–2	2	6	–8	12
–4	4	12	–16	24
5	–5	–15	20	–30
7	–7	–21	28	–42

c

×	–5	6	7	–8
–2	10	–12	–14	16
3	–15	18	21	–24
4	–20	24	28	–32
–5	25	–30	–35	40

7 **a** –6 **b** 4 **c** –3 **d** 75 **e** 24 **f** 8 **g** –2 **h** 6
i –6 **j** 8 **k** –2 **l** –8 **m** –2 **n** –4 **o** 7

8 **a** i 4 **ii** 16 **iii** 9 **iv** 36 **b** Because –ve × –ve = positive and +ve × +ve = positive

9 **a** –2 **b** 2 **c** –14 **d** –4 **e** 26 **f** –10 **g** 4 **h** 3

10 **a** 2×(–5+4) **b** (–2+ –6)×3 **c** 9 – (5 – 2)

Challenge: Algebraic magic squares

A 3*c* **B**

9	–14	11
4	2	0
–7	18	–5

C

–6	–1	–8
–7	–5	–3
–2	–9	–4

Lesson 1.2 Factors and highest common factors (HCF)

Learning objective

- To understand and use highest common factors

Resources and homework

- Pupil Book 2.2, pages 10–13
- Homework Book 2, section 1.2
- Online homework 1.2, questions 1–10

Links to other subjects

- **Design and technology** – to split two different amounts into equal quantities

Key words

- common factor
- factor
- integer
- divisible
- highest common factor (HCF)

Problem solving and reasoning help

- Encourage pupils to list all the factors of both numbers when finding the HCF.
- Question 6 of Exercise 1B in the Pupil Book asks pupils to use the HCF for cancelling down fractions to their lowest terms. It may be worth demonstrating this for **less able** pupils.

Common misconceptions and remediation

- Pupils sometimes confuse factors and multiples. (Tell them that multiples come from multiplying.)

Probing questions

- Two numbers have a HCF of 7 what could they be?
- Convince me that the HCF of 30 and 45 is 15.

Part 1

- Pupils should use whiteboards for their answers, and hold up the boards when requested.
- Ask for examples of: an even number; a multiple of 6; a factor of 12; a prime number; a square number; a number that is a multiple of 3 and 4 at the same time; a triangle number.
- Go around the class each time, checking pupils' answers. If necessary, discuss and define what was required. Emphasise factors, as these will be used in Part 2.
- Feedback suggests that one of the biggest barriers to learning is that pupils do not feel comfortable using mathematical terminology such as: congruent, perimeter, radius. Encourage simple activities to help pupils to remember and work with these words.

Part 2

- Ask pupils to try and write down a number that is a factor of 12 and a factor of 18.
- Write all the suggested numbers on the board. It is likely that all possibilities will be shown (1, 2, 3, 6, plus some that may be incorrect). Make sure that pupils understand the idea, and if any factors are missing ask what is needed to complete the set.
- Ask: 'What is special about 6?' Emphasise that it is the highest common factor or HCF.
- Repeat for the common factors of 30 and 50.
- Now ask for the highest common factor of 16 and 20.
- Repeat for 15 and 30. (5 is a likely answer – make sure pupils realise that the HCF is 15.)
- Repeat for 7 and 9. Ask why the answer is 1.
- Prime numbers should have been defined in Part 1. If not, define prime numbers.
- **Pupils can now do Exercise 1B from Pupil Book 2.2.**

Part 3

- Write numbers on the board (or have prepared cards) such as: 10, 12, 15, 20, 24, 25, 30, 35, 40, 48.
- Ask pupils to pick out one card and to list all the factors.
- Ask pupils to pick out two cards. Ask them for the HCF.
- Alternatively, ask for a card that is the highest common factor of, for example, 15 and 20.

Answers

Exercise 1B

1 **a** 1, 3, 5, 15 **b** 1, 2, 4, 5, 10, 20 **c** 1, 2, 4, 8, 16, 32
d 1, 5, 7, 35 **e** 1, 2, 3, 4, 5, 6, 10, 12, 15, 20, 30, 60

2 **a** 5 **b** 15 **c** 20 **d** 4

3 2, 3, 4, 6, 9, 12, 18

4 **a** 1, 2, 4 **b** 1, 2, 3, 4, 6, 12 **c** 1, 5 **d** 1, 3, 5, 15
e 1, 2, 5, 10 **f** 1, 2, 4, 8 **g** 1, 3 **h** 1, 7

5 **a** 3 **b** 4 **c** 2 **d** 4 **e** 2 **f** 2 **g** 9 **h** 1

6 **a** 20 **b** 9 **c** 4 **d** 28 **e** 15 **f** 36 **g** 50 **h** 40

7 **a** $\frac{2}{3}$ **b** $\frac{2}{3}$ **c** $\frac{3}{5}$ **d** $\frac{2}{5}$ **e** $\frac{3}{8}$ **f** $\frac{4}{5}$ **g** $\frac{9}{20}$ **h** $\frac{8}{9}$

8 Seven groups of nine pupils

Investigation: Tests for divisibility

A It ends in an even number
B The digits will add up to a multiple of 3
C If it ends with a 0, 4 or 8 and the tens digit is 0 or even – or if it ends with a 2 or 6 and the tens digit is odd
D Ends in a 0 or a 5
E If it's even and a multiple of 3
F The digits will add up to a multiple of 9
G **a** 108, 390, 4503, 111111 **b** 108, 220, 580, 12716 **c** 108, 390, 12716 **d** 108

Lesson 1.3 Lowest common multiple (LCM)

Learning objective
- To understand and use lowest common multiples

Resources and homework
- Pupil Book 2.2, pages 13–15
- Homework Book 2, section 1.3
- Online homework 1.3, questions 1–10

Links to other subjects
- **Food technology** – to calculate portions and quantities when preparing food

Key words
- common multiple
- multiple
- lowest common multiple (LCM)

Problem solving and reasoning help
- **PS** questions 8 and 9 of Exercise 1C in the Pupil Book require pupils to work out the LCM of three numbers. Encourage pupils to use the same method they used for two numbers.

Common misconceptions and remediation
- Pupils sometimes confuse factors and multiples. (Say that multiples come from multiplying.)
- Pupils should understand that the multiples of a number are the times table for that number.

Probing questions
- How can we tell if two numbers have no common factors?
- Two numbers have an HCF of 5 and an LCM of 60. What numbers are they?

Part 1
- Pupils should use whiteboards for their answers, and hold up the boards when asked.
- Ask for examples: an odd number; a multiple of 8; a factor of 12; a prime number; a square number; a number that is a multiple of 3 and 4 at the same time; a triangle number.
- Go around the class each time, checking pupils' answers.
- If necessary, discuss and define what was required.
- Particularly emphasise multiples, as these will be used in Part 2.

Part 2
- Still using the whiteboards, ask pupils to write down a number that is a multiple of 3 and 4 (this was asked in part 1).
- On the board, write down all pupils' answers. Ask for more suggestions if many answers are the same.
- Hopefully, 12 will have been suggested. Ask pupils what is special about this. Emphasise that it is the lowest common multiple or LCM.
- Repeat for a common multiple of 4 and 5.
- Ask for the lowest common multiple of 3 and 5.
- Then ask for the lowest common multiple of 4 and 6.
- Many pupils will answer 24, as they will have spotted that previous answers were the product of the two numbers in question.
- Make sure that all pupils understand that in fact 12 is the LCM of 4 and 6.
- If pupils are having trouble at this stage, encourage them to write down the multiples for the two numbers and look for the first common value in each list. For example, for the LCM of 4 and 5, the common value is 20:

  4 8 12 16 **20** 24 28 … 5 10 15 **20** 25 30 35 …
- **Pupils can now do Exercise 1C from Pupil Book 2.2.**

Part 3

- Write numbers on the board (or have prepared cards) such as: 1, 2, 3, 4, 6, 8, 10, 12, 15, 20.
- Ask pupils to pick out one card and then write down the first 10 multiples
- Ask pupils to pick out two cards. Ask for the LCM of both cards.
- Alternatively, ask for a card that is the lowest common multiple of, for example, 5 and 6.

Answers

Exercise 1C

1 **a** 10,4,18,8,72,100 **b** 18,69,81,33,72 **c** 10,65,100 **d** 18,81,72

2 **a** 4,8,12,16,20,24,28,32,36,40 **b** 5,10,15,20,25,30,35,40,45,50
c 8,16,24,32,40,48,56,64,72,80 **d** 15,30,45,60,75,90,105,120,135,150
e 20,40,60,80,100,120,140,160,180,200

3 **a** 40 **b** 20 **c** 30 **d** 40

4 **a** 45 **b** 25 **c** 24 **d** 12 **e** 24 **f** 60 **g** 63 **h** 77

5 **a** 30 **b** 60 **c** 120 **d** 90 **e** 24 **f** 240 **g** 252 **h** 231

6 480

7 112 seconds

8 630cm

9 4 or 5 depending on when they were first all together

Challenge: LCM and HCF

A **a** 2 and 14 **b** 3 and 18 **c** 5 and 60

B **a** **i** 1, 35 **ii** 1, 12 **iii** 1, 22 **b** xy

C **a** **i** 5, 10 **ii** 3, 18 **iii** 4, 20 **b** y

Lesson 1.4 Powers and roots

Learning objective

- To understand and use powers and roots

Resources and homework

- Pupil Book 2.2, pages 16–19
- Intervention Workbook 2, pages 28–29
- Intervention Workbook 3, pages 31–33
- Homework Book 2, section 1.4
- Online homework 1.4, questions 1–10

Links to other subjects

- **Science** – to use formulae in experiments

Key words

- cube
- factor tree
- square
- cube root
- power
- square root

Problem solving and reasoning help

- **PS** question 10 of Exercise 1D in the Pupil Book encourages pupils to spot the patterns arising from question 1.

Common misconceptions and remediation

- Reinforce the fact that the square root of a number can be both positive and negative.
- Another problem is that pupils often think that n^2 is $n \times 2$ or that n^3 is $n \times 3$. Explain clearly that this is not the case.

Probing questions

- Are the following statements always, sometimes or never true?
  - Cubing a number makes it bigger.
  - The square of a number is always positive.
  - You can find the square root of any number.
  - You can find the cube root of any number.
- If sometimes true, precisely when is the statement true and when is it false?

Part 1

- Use a target board such as the one shown.
- Assign values to *a* and *b*. These need to be squares, for example, $a = 1$ and $b = 4$.
- Randomly select pupils and ask them to evaluate the expressions.
- Repeat with other values for *a* and *b*, for example, $a = 4$ and $b = 9$.

a^2	$\sqrt{a}$	$2b^2$	$\sqrt{b}$
$3\sqrt{b}$	$3a^2$	$\frac{\sqrt{a}}{2}$	$2a^2$
b^2	$3\sqrt{b}$	$\frac{a^2}{2}$	$2\sqrt{a}$
$\frac{\sqrt{b}}{2}$	$\frac{b^2}{2}$	$2\sqrt{b}$	$3b^2$

Part 2

- Following on from Part 1, one of the problems asked earlier was 'If $a^2 = 9$, what is *a*?'
- Pupils may only have identified 3. If so, ask for another solution. Obtain the answer: –3.
- Another problem asked earlier was 'What is $\sqrt{9}$?' Is there another answer to this? Again, it is unlikely that –3 will have been given earlier.
- Emphasise that a square root is generally accepted as the positive square root. Then explain the subtle point that the solution to the equation $a^2 = 9$ can be positive or negative.
- Ask pupils what we mean by a^3.
- If $a = 2$, what is a^3? If $a = 3$, what is a^3? If $a = 4$, what is a^3? If $a = 5$, what is a^3?

- Pupils should have the mental skill to work out up to 5^3, but may find 6^3 difficult. You could write out the sequences, for example: 1, 8, 27, 64, 125, 216, 343, 512, 729, 1000.
- If $a^3 = 729$, what is a? It is likely that answers of –9 and 9 will be given. Demonstrate that: $-9 \times -9 \times -9 = -729$. Hence, only one answer is possible for a^3 (= 729).
- Introduce the notation of cube root, $\sqrt[3]{729} = 9$. What is $\sqrt[3]{64}$? What is $\sqrt[3]{125}$?
- Ask pupils what 2^4 means. What is the value of 2^4? (= 16)
- What about 3^5? Calculator may be needed here (= 243).
- **Pupils can now do Exercise 1D from Pupil Book 2.2.**

Part 3

- Here is a quick factual recall test of, for example, squares, cubes, and so on.

1 What is the cube root of 64?
2 What is 1000 as a power of 10?
3 What is 3 cubed?
4 What is the square root of 196?
5 What is the cube root of 1000?
6 What is 5 cubed?
7 What is 1 million as a power of 10?
8 What is –2 squared?
9 If x squared equals x cubed, what is x?
10 What are the values of x if $x^2 = 25$?

Answers

Exercise 1D

1

16	25	36	49	64	81	100
64	125	216	343	512	729	1000

2 **a** 2 **b** 8 **c** 9 **d** 10 **e** 5 **f** 3 **g** 5 **h** 10 **i** 8 **j** 9
3 **a** 169 **b** 2197 **c** 225 **d** 3375 **e** 441 **f** 9261 **g** 1.96
h 5.832 **i** 12.167 **j** 20.25 **k** 1728 **l** 3.375
4 **a** 16 **b** 243 **c** 81 **d** 32 **e** 256 **f** 625 **g** 2401 **h** 512 **i** 128
j 512 **k** 1024 **l** 59049
5 **a** 400 **b** 27000 **c** 125000 **d** 3200000 **e** 4900 **f** 8000000
6 10^4, 10^5, 10^6, 10^7
7 **a** ±6 **b** ±11 **c** ±12 **d** ±1.5 **e** ±14 **f** ±2.4 **g** ±1.6 **h** ±60
8 All answers are 1
9 **a i** 1 **ii** –1 **iii** 1 **iv** –1 **v** 1 **b i** –1 **ii** 1
10 **a** 729 **b** 16^3

Challenge: Squares on a chessboard

204

Lesson 1.5 Prime factors

Learning objectives

- To understand what prime numbers are
  To find the prime numbers of an integer

Resources and homework

- Pupil Book 2.2, pages 19–21
- Intervention Workbook 2, pages 28–29
- Homework Book 2, section 1.5
- Online homework 1.5, questions 1–10

Links to other subjects

- **IT** – to use internet security for key encryption

Key words

- factor tree
- prime factor
- index form
  prime number

Problem solving and reasoning help

- The challenge question at the end of Exercise 1E in the Pupil Book introduces Venn diagrams. Ensure that **less able** pupils understand these before they work on the question.

Common misconceptions and remediation

- Remind pupils to include the multiplication signs when writing a number as a product of its prime factors. (These are often replaced incorrectly by addition signs or commas.)

Probing questions

- Give reasons why none of the following are prime numbers: 4094, 1235, 5121
- How do you go about finding the prime factors of a given number?
- What are the prime factors of 125? 81? 343? What do you notice?
- Can you think of a number that has one repeated prime factor? Or all different prime factors?

Part 1

- Using a target board like the one shown, point to a number and ask a randomly-picked pupil to give the factors of the number.
- Recall the rule for factors – that is, they come in pairs, except for square numbers.
- 1 and the number itself are always factors. Prime numbers only have two factors.

25	36	70	64	75
81	18	50	20	45
30	63	80	92	16
32	15	10	28	60

Part 2

- Ask for the answer to: $2 \times 3 \times 3$ (= 18).
- What about: $2 \times 2 \times 5$ (= 20), $3 \times 5 \times 5$ (= 75), $3 \times 3 \times 7$ (= 63)?
- What can you say about the numbers in the multiplication? They are all prime numbers.
- This is the prime factor form of a number (the number broken down into a product of primes).
- How can we find this if we start with the number 30, for example?
- Explain the tree method: split 30 into a product such as 2×15; then continue splitting any number in the product that is not a prime.
- This can easily be seen in the form of a 'tree'. For example, find the prime factors of 120.
- An alternative is the division method where the number is repeatedly divided by any prime that will go into it exactly.

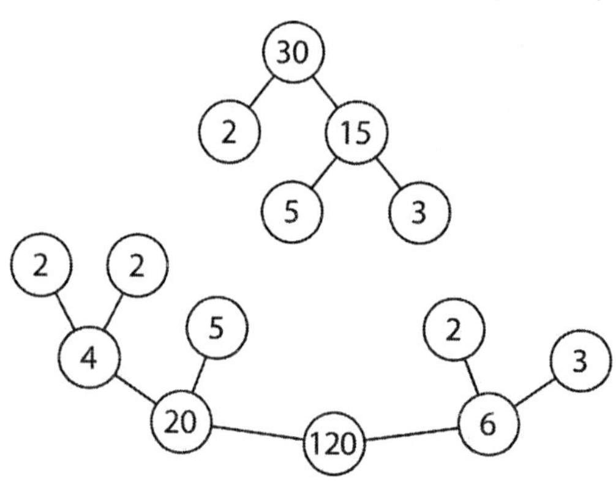

- Demonstrate this with 50. Continue to divide by primes until the answer is 1.

  2|50
  5|25
  5|5
  1

- Repeat with 96 (2 × 2 × 2 × 2 × 2 × 3), 60 (2 × 2 × 3 × 5).
- Now it may be useful to introduce the index notation, that is: 96 = 2^5 × 3, 60 = 2^2 × 3 × 5.
- **Pupils can now do Exercise 1E from Pupil Book 2.2.**

Part 3

- Choose a number, for example, 70.
- Find the factors (1, 2, 5, 7, 10, 14, 35, 70) and the prime factors (2 × 5 × 7).
- Choose another number and repeat, for example, 90.
- The factors are 1, 2, 3, 5, 6, 9, 10, 15, 18, 30, 45, 90, and the prime factors are 2 × 3^2 × 5.
- Ask pupils if they can spot a connection. This is simply that only the prime numbers in the list of factors appear in the prime factors.
- Discuss how to find the HCF and LCM of 70 and 90 (HCF 10, LCM 630).
- Repeat with 48 and 64 (HCF 16, LCM 192).

Answers

Exercise 1E

1 **a** 12 **b** 90 **c** 36 **d** 270 **e** 150
2 **a** 2 × 2 × 2 **b** 2 × 5 **c** 2 × 2 × 2 × 2 **d** 2 × 2 × 5 **e** 2 × 2 × 7
f 2 × 17 **g** 5 × 7 **h** 2 × 2 × 13 **i** 2 × 2 × 3 × 5 **j** 2 × 2 × 3 × 3 ×5
3 **a** 2 × 3 × 7 **b** 3 × 5 × 5 **c** 2 × 2 × 5 × 7 **d** 2 × 5 × 5 × 5 **e** 2 × 2 × 2 × 2 × 2 × 3 × 5
4 2→2, 3→3, 4→2 × 2, 5→5, 6→2 × 3, 7→7, 8→2 × 2 × 2, 9→3 × 3, 10→2 × 5, 11→11, 12→2 × 2 × 3, 13→13, 14→2 × 7, 15→3 × 5, 16→2 × 2 × 2 × 2, 17→17, 18→2 × 3 × 3, 19→19, 20→2 × 2 × 5
5 **a** 2,3,5,7,11,13,17,19 **b** Prime numbers
6 **a** $2^3 \times 5^2$ **b** 2×5^2 **c** $2^3 \times 5^3$ **d** $2^6 \times 5^6$
7 **a** 2 × 5 = 10 **b** 2 × 3 × 5 = 30
8 **a** 2^5 **b** 2^6 **c** 2^7 **d** 2^{10}
9 **a** $2^2 \times 3 \times 5$ **b** $2^2 \times 5^2$ **c** 300 **d** $2^2 \times 3 \times 5^2$
10 420

Challenge: LCM and HCF in Venn diagrams

A **a** 6, 360 **b** 10, 450 **c** 12, 336

B

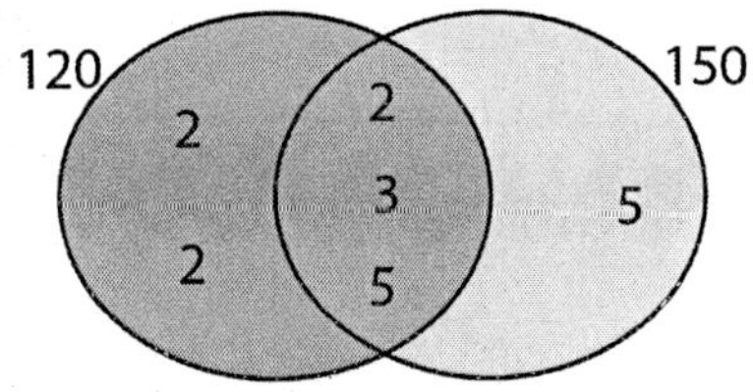

HCF = 30
LCM = 600

C

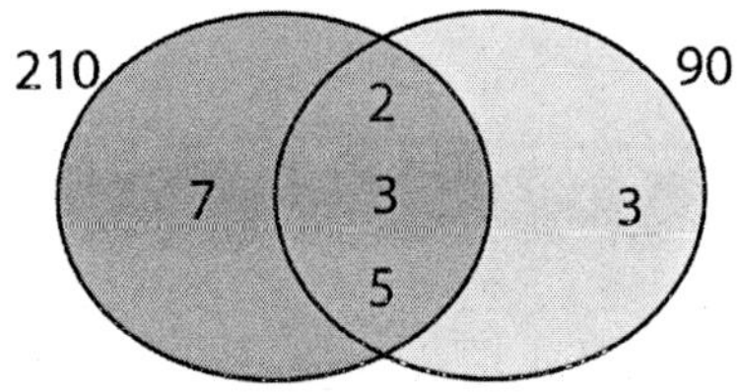

HCF = 30
LCM = 630

Review questions (Pupil Book pages 22–23)

- The review questions will help to determine pupils' abilities with regard to the material within Chapter 1.
- The answers are on the next page of this Teacher Book.

Challenge – Blackpool Tower (Pupil Book pages 24–25)

- This challenge activity encourages pupils to think about a tourist attraction with different facilities and what is involved in running them. The topic could lead to class discussion about environmental issues such as electricity and water usage.
- Look at the information on page 24 of the Pupil Book. Ask pupils some questions relating to the activity. Topics could include:
    - periods of time (when did it open, how long since the foundation stone was laid?)
    - conversion factors (imperial to metric, square metres to square centimetres)
    - data from the internet (search for 'Blackpool Tower').
- Encourage pupils, in small groups or individually, to suggest possible questions. Then they could present the questions to the class for answers. This will be particularly useful if pupils have access to the internet.
- Pupils can now work through questions 1–7 in the Pupil Book.
- Ask pupils to develop questions for another tower they have researched, for example, the Empire State Building.

Answers to Review questions

1 b, d and e

2 **a** 2^4, 5^2, 3^3 **b** 16384

3 For example 2 × –5 and 20 ÷ –2

4 **a** –4, 7 **b** 4, –7 **c** 5, 6

5 50cm

6 **a** 144m **b** 3m

7 **a** 25 × 25 × 36, £225 **b** 49 × 49 × 64, £289

8 **a** 210 **b** 17

9 **a**

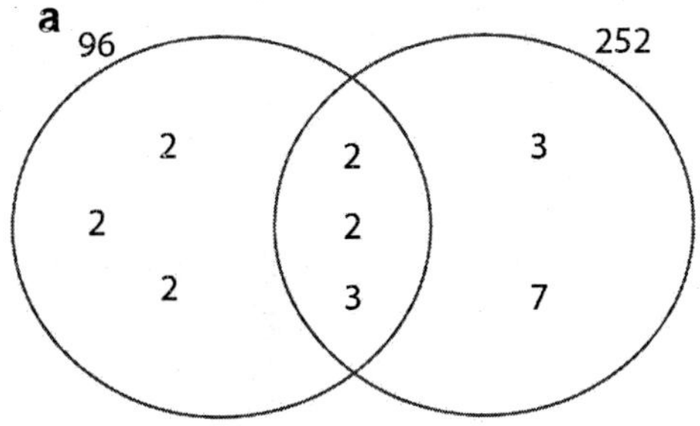

b i 12 **ii** 2016

Answers to Challenge – Blackpool Tower

1 7111 gallons

2 **a** $\frac{1}{40}$ **b** $2\frac{1}{2}$p

3 **a** The Eiffel Tower was cheaper by 61p, at £8.89 **b** 2 times **c** 13 times

4 437 cm²

5 £78 840

6 **a** 190 million **b** 6.57 m

7 No, D = 39.5 km and the Isle of Man is 42 km away

2 Geometry

Learning objectives

- How to identify alternate and corresponding angles
- How to classify quadrilaterals according to their geometrical properties
- How to rotate a shape
- How to translate a shape
- How to construct perpendicular lines and angle bisectors

Prior knowledge

- How to identify parallel and perpendicular lines
- How to calculate angles at a point, angles on a line and opposite angles
- How to calculate angles in a triangle
- How to recognise the symmetry of quadrilaterals
- How to plot coordinates in all four quadrants
- How to use a pair of compasses to draw a circle

Context

- Architects use the properties of angles and shapes to design buildings. Point out to pupils that when looking at buildings one can often see triangles, quadrilaterals and other shapes.
- Understanding angles, lines and shapes and being able to construct them accurately are important in all types of designs. For example, many fabrics and wallpapers use angles, lines and shapes that have been rotated, reflected and repeated to make patterns that look good and are easy to reproduce. When fabrics and wallpaper are printed, manufacturers reproduce the designs by translating them repeatedly. Search for 'frieze patterns' or use the following link as an example of how transformations have been used to create designs throughout history:
  http://plato.acadiau.ca/courses/educ/reid/Geometry/Symmetry/frieze.html

Discussion points

- What type of angle is this? How do you know?
- How can you use what you know about acute and obtuse angles to check your measurement?
- Is your measurement sensible? Why?
- Show visual example of angles and shapes, some of which have been drawn or labelled incorrectly. Ask pupils to identify which examples have been drawn incorrectly using angle facts and characteristics of shapes.
- An isosceles triangle has one angle of 35°. Is this enough information to know the other two angles? Why?

Associated Collins ICT resources

- Chapter 2 interactive activities on Collins Connect online platform
- *Pencils* and *Miura Folding* Wonder of Maths on Collins Connect online platform

Curriculum references

Reason mathematically

- Begin to reason deductively in geometry, number and algebra, including using geometrical constructions

Solve problems

- Develop their mathematical knowledge, in part through solving problems and evaluating the outcomes, including multi-step problems

Geometry and measures

- Derive and use the standard ruler and compass constructions (perpendicular bisector of a line segment, constructing a perpendicular to a given line from/at a given point, bisecting a given angle)
- Describe, sketch and draw using conventional terms and notations: points, lines, parallel lines, perpendicular lines, right angles, regular polygons, and other polygons that are reflectively and rotationally symmetric
- Derive and illustrate properties of triangles, quadrilaterals, circles, and other plane figures
- Identify properties of, and describe the results of, translations, rotations and reflections applied to given figures
- Apply the properties of angles at a point, angles at a point on a straight line, vertically opposite angles
- Understand and use the relationship between parallel lines and alternate and corresponding angles

Fast-track for classes following a 2-year scheme of work

- Much of the material in this chapter will be familiar to pupils. Use the activities and challenges at the end of each exercise to check pupils' understanding. If you feel that pupils' understanding is secure, move straight on to Lesson 2.5.

Lesson 2.1 Angles in parallel lines

Learning objective
- To calculate angles in parallel lines

Resources and homework
- Pupil Book 2.2, pages 27–32
- Homework Book 2, section 2.1
- Online homework 2.1, questions 1–10

Links to other subjects
- Design and technology – to construct models of designs

Key words
- allied angles
- corresponding angles
- alternate angles
- transversal

Problem solving and reasoning help

- **MR** questions 3, 4, 5, 7 and 9 of Exercise 2A in the Pupil Book encourage pupils to explain and justify their solutions. This is very important as these are the early steps towards mathematical proof. Provide lots of opportunity for discussion and encourage pupils to reflect on and extend the responses of other pupils. Make sure pupils understand that correcting and adapting previous statements is part of the learning process. Use activities such as reading images to encourage pupils to explore ideas collaboratively and to persevere when trying to come to a mathematical justification.

Common misconceptions and remediation

- Pupils often assume that because something looks correct that it is true. They need to understand the importance of correct mathematical notation, for example, to identify parallel lines, and the need for rigorous mathematical proof.

Probing questions

- Explain why a triangle cannot have two parallel sides.

Part 1

- Use a target board, as shown. You could repeat this activity using different sets of numbers or rules, but the theme is supplementary and complementary angles (which sum to 180° and 90°, respectively).
- Ask pupils to choose a number from the target board, and subtract it from 180.
- Ask pupils to find two numbers that add up to 180. Then ask pupils to add two numbers together to give a total less than 180, and subtract the result from 180.
- Ask pupils to find two numbers that add up to 90; then three numbers that add up to 180.

90	70	50	30	60	58	73
45	105	32	17	127	15	165
63	87	25	120	148	20	3
163	135	75	110	130	65	40

Part 2

- Draw a letter Z (with a rotational symmetry order of two) on a piece of card. Tell the class that the main activity involves finding angles. Ask pupils what happens to the letter Z when the card is rotated through 180°, and to be specific about what happens to the two angles.

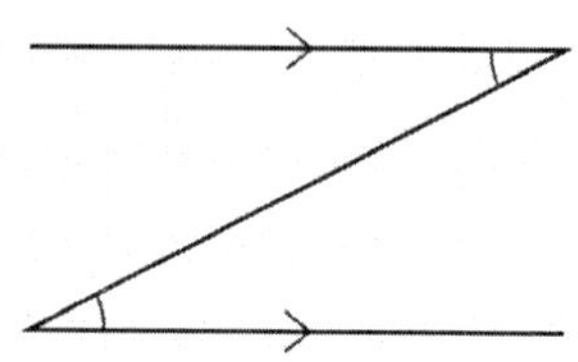

- Demonstrate, by using an acetate film or sheet with a tracing of the letter on it, that the top and bottom angles are equal.
- Now ask the class what happens to the two angles if a letter Z is drawn without the top and bottom lines being parallel.
- You could use Geo-Strips to demonstrate this. Ensure that the class recognise that if the lines are parallel, then the angles are equal; if they are not parallel, then the angles differ.
- Stress to pupils that this may help them to visualise these types of angles but that the correct mathematical name is *alternate angles*.
- Now use Geo-Strips or a letter F drawn on a piece of card or the board to establish the rules for *corresponding angles*.
- Again stress to pupils that this may help them visualise these types of angles but that the correct mathematical name is corresponding angles.
- **Pupils can now do Exercise 2A from Pupil Book 2.2.**

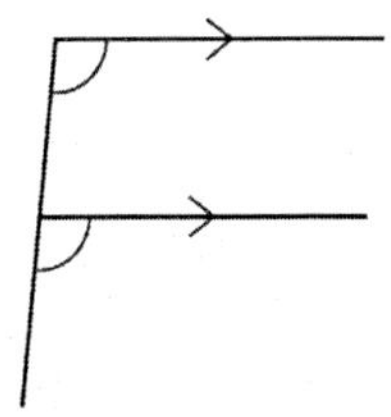

Part 3

- Summarise the lesson by pointing out that in questions with many parallel lines, often many of the angles will be the same and many others will be the supplement of the first angle.
- Remind pupils that they need to be able to work out angles and use the correct terminology.

Answers

Exercise 2A

1 **a** *e* **b** *f* **c** *g* **d** *h* **e** *d* **f** *c*

2 **a** *a* = 70° **b** *b* = 125° **c** *c* = 160° **d** *d* = 48°
e *e* = 75° **f** *f* = 57° **g** *g* = 121° **h** *h* = 34°

3 b, the other two show a pair of alternate angles

4 c, the other two show a pair of corresponding angles

5 a, the other two show a pair of allied angles

6 **a** *a*= 50° (alternate angles) **b** *b*= 62° (corresponding angles)
c *c*= 108° (alternate angles) **d** *d*= 41° (alternate angles)
e *e*= 124° (corresponding angles) **f** *f*= 63° (corresponding angles)

7 **a** *a*= 122° (angles on a line), *b* = 58° (corresponding angles or allied angles), *c* = 58° (opposite angles)
b *d* = 60° (opposite angles), *e* = 60° (corresponding angles), *f* = 120° (allied angles)

8 *x* and *y* are allied angles and add up to 180°

9

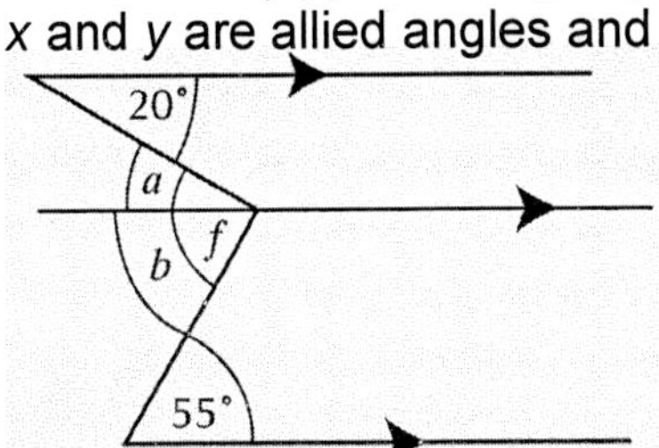

Extend the middle parallel line to split angle *f* into 2 angles *a* and *b*
a = 20° (alternate angles are equal)
b = 55° (alternate angles are equal)
So *f* = 20° + 55° = 75°

Mathematical reasoning: More about parallel lines

Note: other reasons are possible

A *a* = 85° (corresponding angles are equal), *b* = 75° (corresponding angles are equal), *c* = 85° (alternate angles are equal)

B *d* = 90° (alternate angles are equal), *e* = 42° (alternate angles are equal)

C *f* = 65° (corresponding angles are equal), *g* = 115° (angles on a line add up to 180°)
h = 65° (corresponding angles are equal), *i* = 115° (allied angles add up to 180°)

D *j* = 98° (angles on a line add up to 180°), *k* = 33° (alternate angles are equal),
l = 147° (angles on a line add up to 180°), *m* = 98° (allied angles add up to 180°)

E *n* = 35° (angles on a line add up to 180°, then angles in a triangle add up to 180°),
o = 83° (corresponding angles are equal, then angles on a line add up to 180°),
p = 118° (opposite angles are equal), *q* = 118° (corresponding angles are equal)

Lesson 2.2 The geometric properties of quadrilaterals

Learning objectives

- To know the geometric properties of quadrilaterals

Links to other subjects

- **Design and technology** – to construct models of designs
- **Geography** – to identify locations on maps by descriptions of the 2D representation of man-made objects

Resources and homework

- Pupil Book 2.2, pages 33–35
- Homework Book 2, section 2.2
- Online homework 2.2, questions 1–10

Key words

- geometric properties

Problem solving and reasoning help

- The investigation at the end of Exercise 2B encourages pupils to be methodical when exploring geometrical problems. Give pupils the opportunity to identify the transferable process skills involved.

Common misconceptions and remediation

- Pupils often assume that because something looks correct that it is true. They need to understand the importance of correct mathematical notation, for example, to identify parallel lines, and the need for rigorous mathematical proof.

Probing questions

- Use ‘sorting cards’, for example: regular polygon, irregular polygon, no lines of symmetry, at least one line of symmetry, no right angles, at least one right angle. Pupils should choose a card; try to find a matching shape, and explain why she or he chose that particular shape.

Part 1

- This is mainly revision of quadrilaterals. Ask the class to imagine a quadrilateral that has only two lines of symmetry: Ask a pupil to draw shape on the board with its name; ask another pupil to draw another possible example (rectangle and rhombus).
- Tell the class that the shape now has equal sides, and ask which of the two quadrilaterals satisfies both conditions (the rhombus).
- Repeat this activity using different quadrilaterals.

Part 2

- Explain to pupils that they will now look at the properties of the quadrilaterals with which they are familiar. Draw each shape shown on the board. Working in pairs, ask the class to describe as many mathematical properties of each shape as they can. Write all the properties below each shape and ask pupils to copy these into their books.

Square

- Four equal sides
- Four right angles
- Opposite sides parallel
- Diagonals bisect each other at right angles
- Four lines of symmetry
- Rotational symmetry of order four

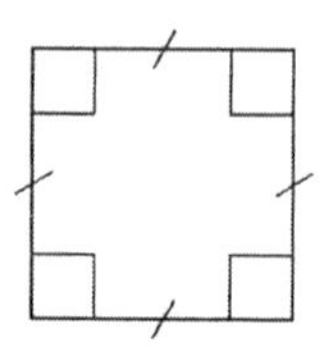

Rhombus

- Four equal sides
- Two pairs of equal angles
- Opposite sides parallel
- Diagonals bisect each other at right angles
- Two lines of symmetry
- Rotational symmetry of order two

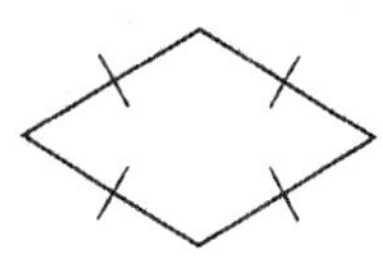

Rectangle

- Two pairs of equal sides
- Four right angles
- Opposite sides parallel
- Diagonals bisect each other
- Two lines of symmetry
- Rotational symmetry of order two

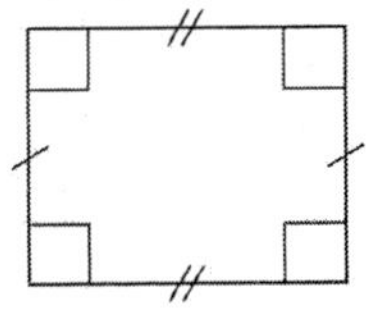

Parallelogram

- Two pairs of equal sides
- Two pairs of equal angles
- Opposite sides parallel
- Diagonals bisect each other
- No lines of symmetry
- Rotational symmetry of order two

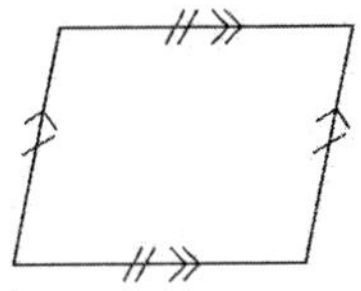

Trapezium

- One pair of parallel sides
- Some trapeziums have one line of symmetry

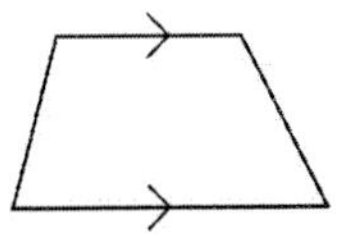

Kite

- Two pairs of adjacent sides of equal length
- One pair of equal angles
- Diagonals intersect at right angles
- One line of symmetry

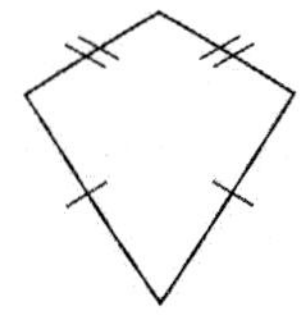

Arrowhead or Delta

- Two pairs of adjacent sides of equal length
- One pair of equal angles
- Diagonals intersect at right angles outside the shape
- One line of symmetry

- **Pupils can now do Exercise 2B from Pupil Book 2.2.**

Part 3

- Ask individuals to draw one of the quadrilaterals that they have met during the lesson on the board. Then they should explain all the mathematical properties of the chosen shape. Ask pupils to provide formative feedback on the quality of the explanation, based on a set of agreed criteria such as clarity, accuracy and use of correct mathematical language.

Answers

Exercise 2B

1

No lines of symmetry	**One line of symmetry**	**Two lines of symmetry**	**Four lines of symmetry**
Parallelogram	Kite	Rectangle	Square
Trapezium	Arrowhead	Rhombus	

2

Rotational symmetry of order one	**Rotational symmetry of order two**	**Rotational symmetry of order four**
Kite	Rectangle	Square
Arrowhead	Parallelogram	
Trapezium	Rhombus	

3 Rectangle

4 Parallelogram

5 Wrong, it could be a rhombus

6 Wrong, it could be a parallelogram or a rhombus

7 Parallelogram, rhombus

8 Use the line of symmetry AC on the kite.
So triangle ADC is identical to triangle ABC.
So $\angle p = \angle q$.

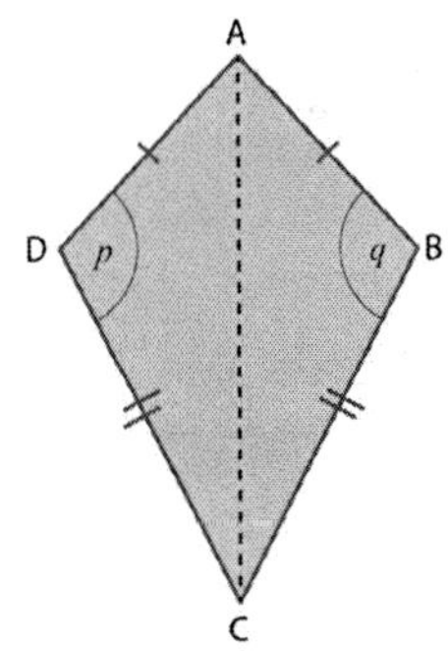

Investigation: Rectangles into squares

For example, for a 4 by 2 rectangle, 3 different ways

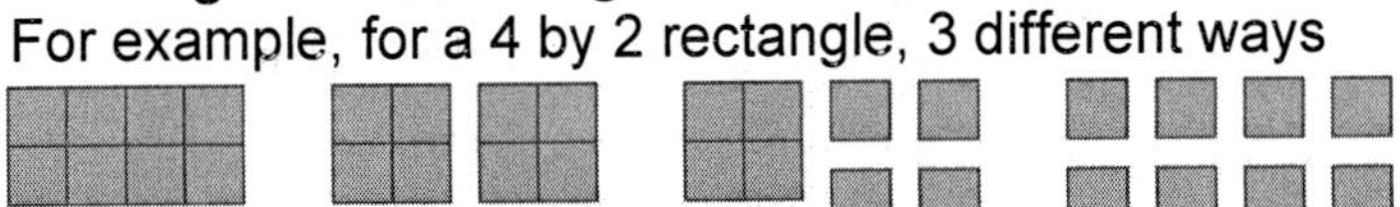

4 by 2

For example, for a 4 by 3 rectangle, 4 different ways

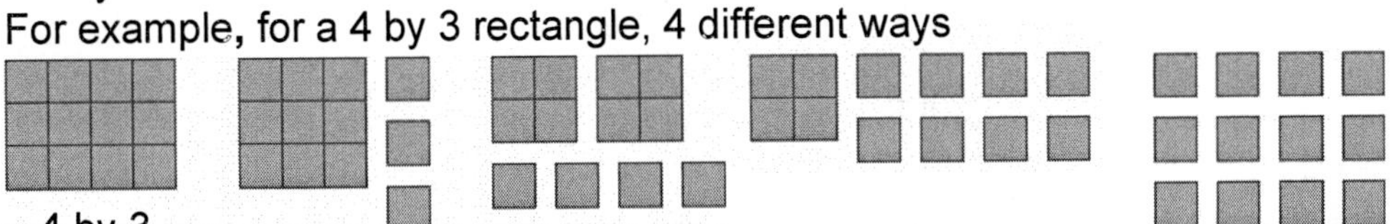

4 by 3

Lesson 2.3 Rotations

Learning objective
- To understand how to rotate a shape

Resources and homework
- Pupil Book 2.2, pages 36–39
- Intervention Workbook 2, pages 53–55
- Homework Book 2, section 2.3
- Online homework 2.3, questions 1–10

Links to other subjects
- **Geography** – to plan towns

Key words
- angle of rotation
- centre of rotation
- direction of rotation
- rotation

Problem solving and reasoning help
- Do an internet search for 'GeoGebraTube – centre of rotation' to help pupils visualise the investigation at the end of Exercise 2C in the Pupil Book, or use this link: **http://www.geogebratube.org/material/show/id/8722**.

Common misconceptions and remediation
- Pupils struggle to visualise rotations. Give them plenty of practice and if possible use active geometry packages such as GeoGebra to help them: **http://www.geogebra.org/cms/en/** You could also use readymade examples on GeoGebra Tube, for example: **http://www.geogebratube.org/material/show/id/2163**

Probing questions
- What properties do you need to know about a quadrilateral to be sure it is a …?

Part 1
- Imagine a square. Now imagine another square, exactly the same size, which touches the first one along all of one of its sides. What shape are you thinking of? (A rectangle)
- Imagine an equilateral triangle. Now imagine another equilateral triangle, exactly the same size, which touches the first one along all of one of its sides. What shape are you thinking of? (A rhombus)

Part 2
- Remind the class about the different transformations they met in Year 7. Draw the following diagrams on the board or on an OHT:

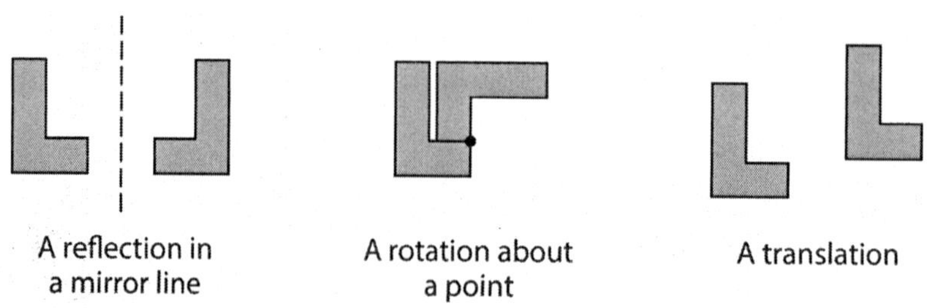

- Ask the class to describe what happens to the shapes after any of the three transformations. They should remember that the object and the image are the same shape and size.
- Say that in this lesson pupils will focus on rotations of the shapes they met in the last lesson.
- **Pupils can now do Exercise 2C from Pupil Book 2.2.** (They may need tracing paper.)

Part 3

- Ask pupils to write a short text or email to a friend to outline the key points they need to know when rotating a shape.
- Pupils could role-play the two sides of the conversation.

Answers

Exercise 2C

1 **a**

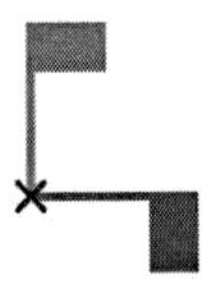

b

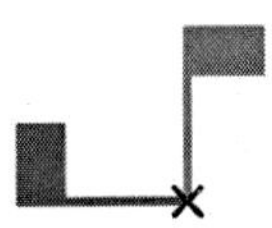

c

2 **a**

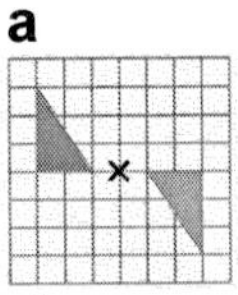

b

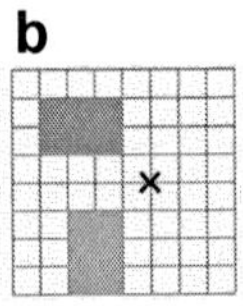

c

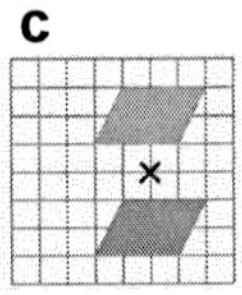

d

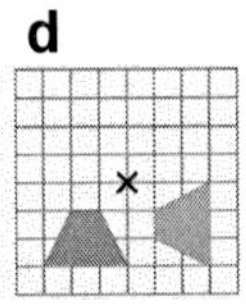

3 **i a**

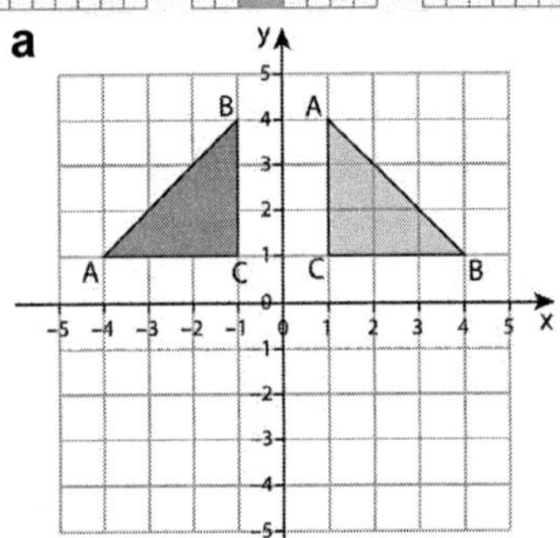

b A(1, 4), B(4, 1), C(1, 1)
c A'(−4, 1), B'(−1, 4), C'(−1, 1)

ii a

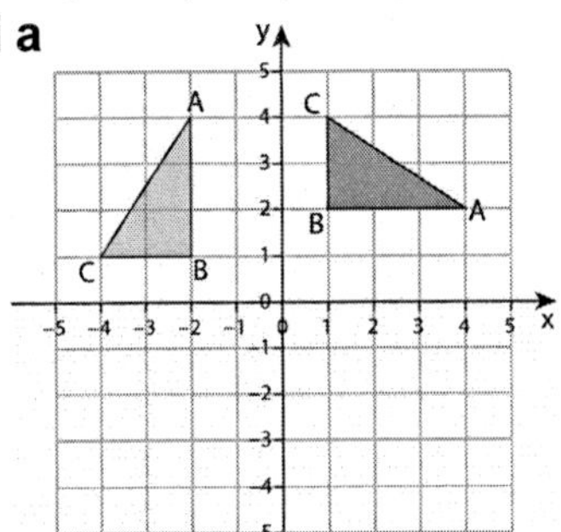

b A(−2, 4), B(−2, 1), C(−4, 1)
c A'(4, 2), B'(1, 2), C'(1, 4)

iii a

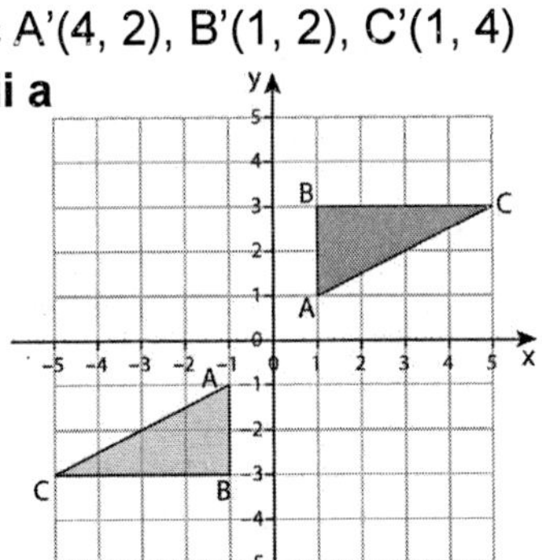

b A(−1, −1), B(−1, −3), C(−5, −3)
c A'(1, 1), B'(1, 3), C'(5, 3)

iv a

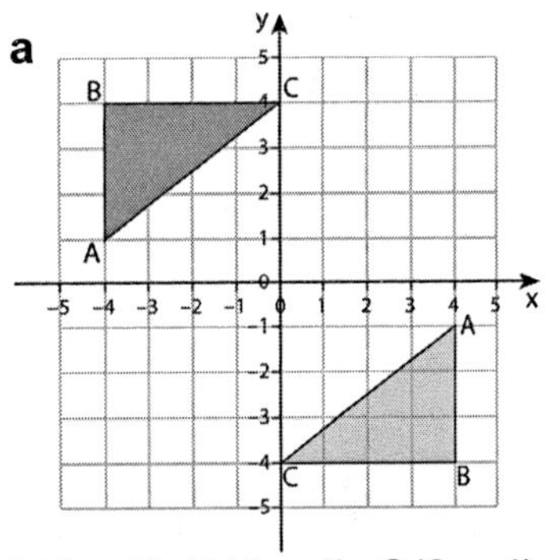

b A(4, −1), B(4, −4), C(0, −4)
c A'(−4, 1), B'(−4, 4), C'(0, 4)

4 **a**

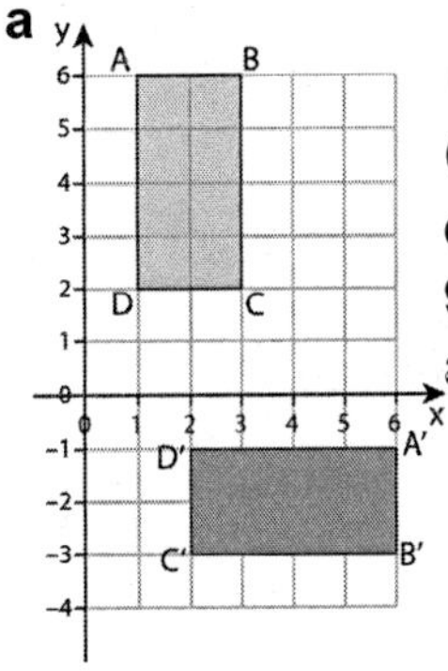

b A'(6, −1), B'(6, −3), C'(2, −3), D'(2, −1)
c A rotation through 90° anticlockwise about the origin O

5 **a**

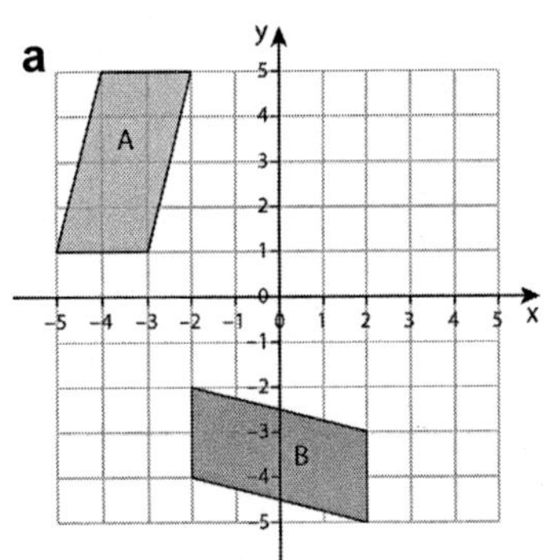

b A rotation through 90° clockwise about the point (2, 1) or a rotation through 270° anticlockwise about the point (2, 1)

Challenge: Finding the centre of rotation

A A rotation of 90° anticlockwise (or 270° clockwise) about the point (2, 0)

B A rotation of 180° clockwise (or anticlockwise) about the point (1, 0)

C A rotation of 90° clockwise (or 270° anticlockwise) about the point (1, −1)

Lesson 2.4 Translations

Learning objective

- To understand how to translate a shape

Resources and homework

- Pupil Book 2.2, pages 40–42
- Intervention Workbook 1, pages 44–45
- Homework Book 2, section 2.4
- Online homework 2.4, questions 1–10

Links to other subjects

- **Design and technology** – to create frieze patterns
- **Physical Education** – to analyse patterns of movement, for example, in dance

Key words

- translate
- translation

Problem solving and reasoning help

- The investigation at the end of Exercise 2D in the Pupil Book encourages pupils to be methodical when exploring geometrical problems. Give pupils the opportunity to identify the transferable process skills involved. You could refer back to the 'frieze patterns' link that was suggested in the opening pages of this chapter.

Common misconceptions and remediation

- A sound understanding of coordinates in all four quadrants will help pupils to understand translations. Physical demonstrations will also help the pupils who may struggle with this.

Probing questions

- What translations can you find in patterns in, for example, flooring, tiling, wallpaper, wrapping paper?
- How would you describe this translation precisely?

Part 1

- Draw a large 'T' in the centre of the board.
- Ask a pupil to draw another 'T' anywhere on the board.
- Ask the class which 'T' is the object and which 'T' is the image. Ask them which transformation could have mapped the object to the image?
- Repeat the activity by allowing pupils to draw an image of 'T' in different orientations.

Part 2

- Remind pupils of the three types of transformations they met at the start of the last lesson.
- Tell pupils that in this lesson they will focus on translations. Say that sometimes they will be given an object and a translation and asked to find the image.
- Alternatively, they may be given an object and an image and asked to describe the translation.
- Work through examples 7 and 8 on page 40 of the Pupil Book to demonstrate each type of question.
- **Pupils can now do Exercise 2D from Pupil Book 2.2.**

Part 3

- Ask individual pupils to explain the following terms: a transformation, a reflection, a rotation, a translation and a combined transformation.

Answers

Exercise 2D

1 **a** 6 units right

b 3 units right and 3 units down

c 6 units down

d 7 units right and 5 units down

e 6 units left and 6 units down

f 4 units right and 2 units down

g 7 units right and 1 unit up

h 7 units left and 5 units up

2 **a** A(1, 4), B(4, 2), C(1, 2)

b and **d**

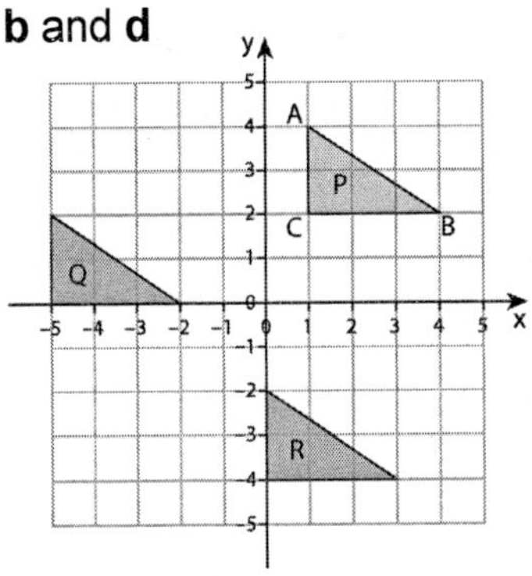

c (–5, 2), (–2, 0), (–5, 0)

e (0, –2), (3, –4), (0, –4)

f 1 unit right and 6 units up

3 **a**

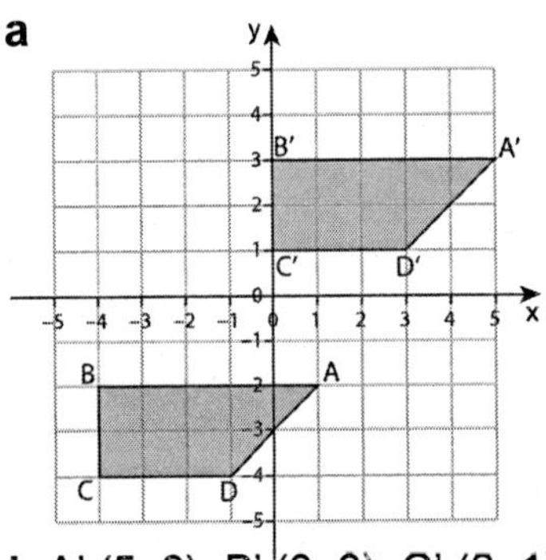

b A' (5, 3), B' (0, 3), C' (0, 1), D' (3, 1)

c Only the position has changed. The size and orientation of the trapezium have stayed the same.

4 **a** A (3, −1), B (2, −3), C (3, −5), D (4, −3)

b and **d**

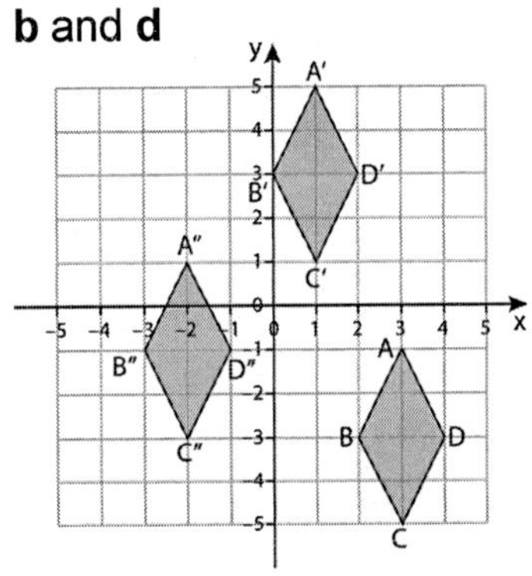

c A' (1, 5), B' (0, 3), C' (1, 1), D' (2, 3)

e A" (−2, 1), B" (−3, −1), C" (−2, −3), D" (−1, −1)

f Translation 5 units right and 2 units down

5 **a** and **b**

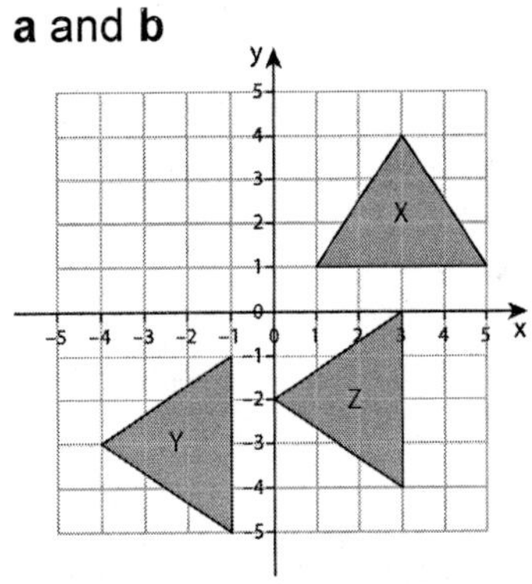

c Translate triangle Z 4 units left and 1 unit down back to triangle Y and then rotate triangle Y through 90° clockwise about the origin O back to triangle X.

Investigation: Dotty translations

Check pupils' answers.

Lesson 2.5 Constructions

Learning objectives

- To construct the mid-point and the perpendicular bisector of a line
- To construct an angle bisector

Resources and homework

- Pupil Book 2.2, pages 43–45
- Homework Book 2, section 2.5

Links to other subjects

- **Design and technology** – to construct plans of 3D objects

Key words

- angle bisector
- construction
- bisect
- perpendicular bisector

Problem solving and reasoning help

- The activity at the end of Exercise 2E in the Pupil Book requires pupils to apply and extend their understanding to a less familiar context. This will provide an opportunity for you to assess whether they really understand what they are doing.

Common misconceptions and remediation

- Pupils are often not precise enough when doing constructions in mathematics. Give them the opportunity to assess the errors in exemplars and explain how they can avoid these errors. Use dynamic geometry software to support pupils. Make sure they understand each step, or they will fail to apply them in more complex, or less familiar contexts.

Probing questions

- Produce your own brief step-by-step guide to constructing a perpendicular bisector to a straight line.
- Produce your own brief step-by-step guide to constructing an angle bisector.

Part 1

- Write on the board, '180 DIAT'. Explain to the class that this is an acronym for 180 degrees in a triangle.
- Ask the class to solve the following geometric acronyms:
  - 90 DIARA (90 degrees in a right angle)
  - 180 DOASL (180 degrees on a straight line)
  - 360 DIACT (360 degrees in a complete turn)
  - 3 SIAT (three sides in a triangle)
  - 4 SIAQ (four sides in a quadrilateral)
  - Ask the class to make up their own geometric acronyms.

Part 2

- For this lesson pupils will need a sharp pencil, a ruler, a pair of compasses and a protractor.
- Show the class how to complete the four constructions in the examples below.
- Explain that these constructions are very accurate when done with a sharp pencil, which is why they are used in other subjects such as Design and technology.

Example 1: *To construct the mid-point and the perpendicular bisector of the line AB*

- Draw a line segment AB of any length. Set compasses to any radius greater than half the length of AB. Draw two arcs, with the centre at A, above and below AB.

- With compasses set at the same radius, draw two arcs with the centre at B, to intersect the first two arcs at C and D. Join C and D to intersect AB at X. X is the mid-point of the line AB.
- The line CD is the perpendicular bisector of the line AB.

Example 2: *To construct the bisector of the angle ABC*

- Draw an ∠ABC of any size. Set compasses to any radius and, with the centre at B, draw an arc to intersect BC at X and AB at Y.
- With compasses set to any radius, draw two arcs with the centres at X and Y, to intersect at Z. Join BZ. BZ is the bisector of the ∠ABC. The ∠ABZ = ∠CBZ.

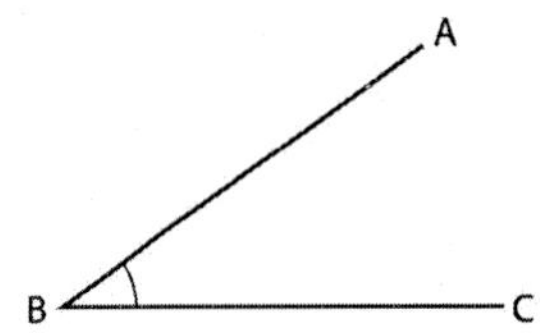

Example 3: *To construct the perpendicular from a point P to a line segment AB:*

- Set compasses to any suitable radius and draw arcs from P to intersect AB at X and Y.
- With compasses set at the same radius, draw arcs with the centres at X and Y to intersect at Z below AB. Join PZ. PZ is perpendicular to AB.

Example 4: *To construct the perpendicular from a point Q on a line segment XY*

- Set compasses to a radius that is less than half the length of XY. With the centre at Q, draw two arcs on either side of Q to intersect XY at A and B. (Line XY may need to be extended).
- Set compasses to a radius that is greater than half the length of XY and, with the centres at A and B, draw arcs above and below XY to intersect at C and D.
- Join CD. CD is the perpendicular from the point Q.
- **Pupils can now do Exercise 2E from Pupil Book 2.2.**

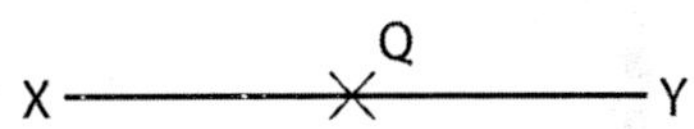

Part 3

- Briefly summarise the four constructions that pupils completed during the lesson.

Answers

Exercise 2E

Check pupils' drawings.

Activity: Construct a line parallel to a given line and passing through a given point

Check pupils' drawings.

Review questions (Pupil Book pages 46–47)

- The review questions will help to determine pupils' abilities with regard to the material within Chapter 2.
- These questions also draw on the mathematics covered in earlier chapters of the book to encourage pupils to make links between different topics.
- The answers are on the following page.

Challenge – More constructions (Pupil Book pages 48–49)

- This challenge gives pupils the opportunity to extend their learning to more complex constructions. They need to be able to reproduce a set of instructions that extend what they have already done during Lesson 2.5 and in Exercise 2E of the Pupil Book.
- This activity is designed to increase pupils' skills for independent work as well as extending the content of Lesson 2.5. Pupils should work in pairs or small groups. Encourage them to work through the examples carefully before asking any questions.
- As a warm-up to this activity, you could ask pupils to summarise what they have learnt during the lesson and say that this activity requires them to build on this knowledge.
- **More able** pupils could develop this further by exploring 'geometric constructions' on the internet. For example: **http://www.mathsisfun.com/geometry/constructions.html**

Answers to Review questions

1 **a** 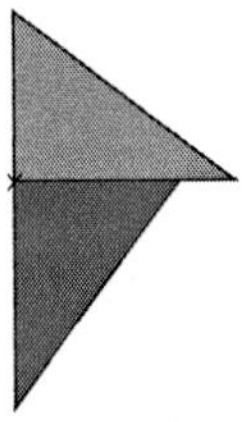 **b** 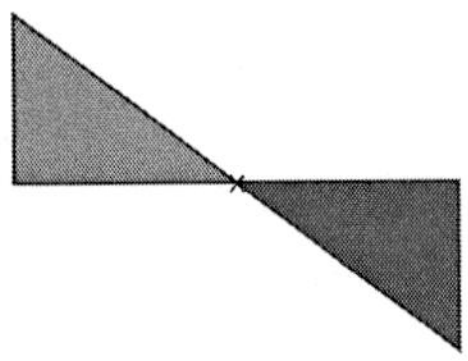 **c** 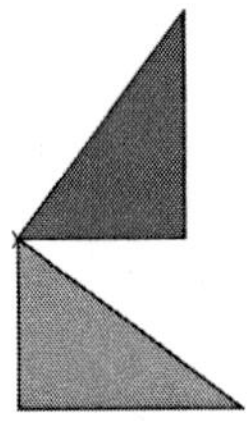

2 **a** and **b**

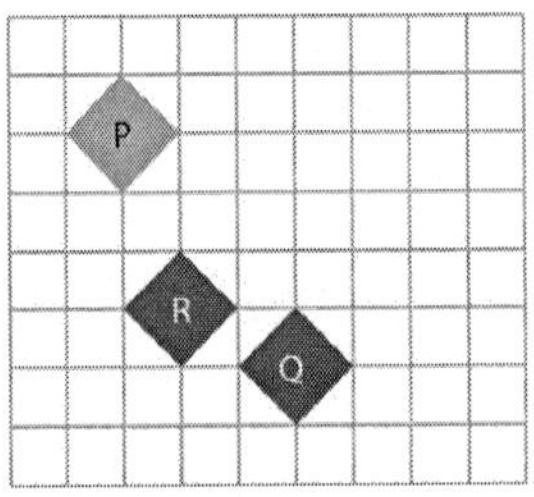

c 1 unit left and 3 units up

3 **a** $a = 112°$ (alternate angles are equal) **b** $b = 56°$ (corresponding angles are equal)
c $c = 43°$ (allied angles add up to 180°) **d** $d = 98°$ (corresponding angles are equal)

4 **a** ∠CBD and∠ABH (alternate angles)
b 50° (the two other angles in the isosceles triangle are equal)
c 95° (allied angles add up to 180°)

5 **a** 110° (allied angles add up to 180°)
b $360° - 3 \times 70° = 150°$, $m = 150° \div 3 = 50°$ or $360° \div 3 = 120°$, $m = 120° - 70° = 50°$

6 **a** A(1, 4), B(3, 4), C(3, 1)
b

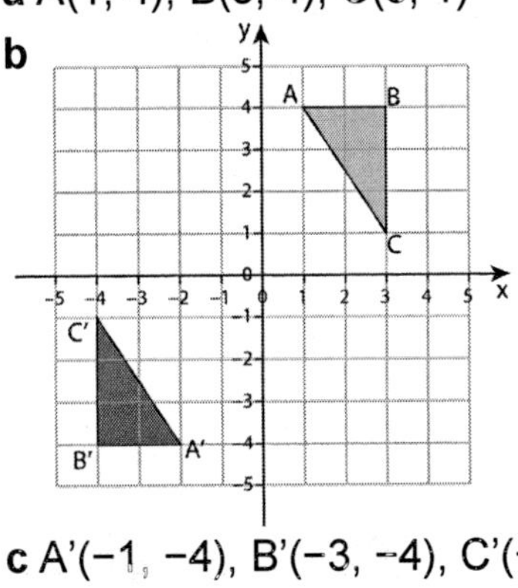

c A'(−1, −4), B'(−3, −4), C'(−3, −1)
d They are the same as triangle ABC, but with a minus sign for each coordinate.
e P'(−2, −1), Q'(−3, −5), R'(−4, −1)

Answers to Challenge – More constructions

Check pupils' constructions.

3 Probability

Learning objectives

- How to work with a probability scale
- How to recognise mutually exclusive and non-exclusive outcomes and events
- How to work out probabilities using sample spaces and Venn diagrams where necessary
- How to use experimental probability to make predictions

Prior knowledge

- What chance and probability are
- How to collect data from a simple experiment
- How to record data in a table or chart

Context

- This chapter builds on previous knowledge of probability and extends this, first to see how probability is applied differently to theory and experiments, and then to being able to compare the two results critically. Pupils should be interested in the 'It's a tie' probability challenge at this link: **http://nrich.maths.org/5516**

Discussion points

- You spin a coin 50 times and count the number of times you get 'heads'. If your friend decided to spin the coin 1000 times, who is most likely to be closer to getting an equal number of heads and tails? Why?

Associated Collins ICT resources

- Chapter 3 interactive activities on Collins Connect online platform
- *Predictions* Wonder of Maths on Collins Connect online platform

Curriculum references

Develop fluency

- Use language and properties precisely to analyse probability and statistics

Reason mathematically

- Explore what can and cannot be inferred in statistical and probabilistic settings, and begin to express their arguments formally

Probability

- Record, describe and analyse the frequency of outcomes of simple probability experiments involving randomness, fairness, equally and unequally likely outcomes, using appropriate language and the 0–1 probability scale
- Understand that the probabilities of all possible outcomes sum to 1
- Enumerate sets and unions/intersections of sets systematically, using tables, grids and Venn diagrams

- Generate theoretical sample spaces for single and combined events with equally likely, mutually exclusive outcomes and use these to calculate theoretical probabilities

Fast-track for classes following a 2-year scheme of work

- Much of the material in this chapter will be new. If pupils are familiar with Lesson 3.1 from Year 7, they can move on to the activity question at the end of Exercise 3A.

Lesson 3.1 Probability scales

Learning objective

- To use a probability scale to represent a chance

Resources and homework

- Pupil Book 2.2, pages 51–55
- Intervention Workbook 3, pages 61–63
- Homework Book 2, section 3.1
- Online homework 3.1, questions 1–10

Links to other subjects

- **Science** – to determine the likelihood of specific experimental outcomes

Key words

- equally likely
- outcome
- probability scale
- event
- probability

Problem solving and reasoning help

- Part b of question 5 of Exercise 3A in the Pupil Book draws on pupils' knowledge of equivalent fractions to solve the problem. The activity question at the end of the exercise requires pupils to have some basic prior knowledge about setting up formulae in spreadsheets. **Less able** pupils may need extra guidance.

Common misconceptions and remediation

- The language of probability is often used in everyday terms, and pupils cannot always tell the difference between the informal use of the language and the more precise use in mathematics. Encourage pupils to discuss and assess the use of the language of probability, for example, in the media. Suggest they think about: How accurate, or not, is the language that is being used?

Probing questions

- Why is it helpful to know that the sum of probabilities of all mutually exclusive events is one? Give me an example of how you have used this when working on probability problems.

Part 1

- Write 10 numbers on the board. Tell the class to try to memorise as many of these numbers as possible. Give pupils about 15 seconds.
- Now cover the numbers and see how many the pupils can recall. This could be a timed exercise, which you could repeat in another lesson.
- Now write 10 fractions and/or decimals on the board.
- Again, see how many the pupils can recall after, for example, 20 seconds.
- This time, return to the 10 fractions and/or decimals and ask the class what is needed to make each one add up to 1. Write the correct answer on the board. To help pupils, you could use a number line drawn on the board or a 'counting stick' with 10 divisions marked on. Mark it at one end with 0 and at the other end with 1.

- Tell pupils that they now have 30 seconds to memorise the answers.

Part 2

- Ask the class to tell you some events that are opposite or exclusive of each other. You may need to provide an example such as: red playing card, black playing card.

- Now tell pupils an event and ask what the opposite is, for example: even number, vowel, wears glasses, does not wear glasses.
- Now ask one pupil to give you a probability as a decimal and ask another pupil for the probability of that event not happening. You could repeat this several times.
- Then you could repeat this with simple fractions.
- In pairs, ask pupils to test each other by making up their own probabilities in turn, and writing them in two columns in their books.
- Introduce the idea that we say that the probability of something happening is *p*, so the expression for the probability of it *not* happening is 1 – *p*.
- Remind the class of the method used to calculate probabilities (introduced in Year 7):

  $$\text{The probability of an event } (p) = \frac{\text{Number of outcomes in the event}}{\text{Total number of all possible outcomes}}$$

- **Pupils can now do Exercise 3A from Pupil Book 2.2.**

Part 3

- Two coins are thrown at the same time. There are four possible outcomes: heads/heads; heads/tails; tails/heads; tails/tails.
- The probability of throwing tails/tails is 'one in four', or $\frac{1}{4}$. What is the probability of not throwing two tails?
- Extend this to: three coins, four coins, and five coins.

Answers

Exercise 3A

1 **a** $\frac{25}{50} = \frac{1}{2}$ **b** $\frac{5}{50} = \frac{1}{10}$ **c** $\frac{14}{50} = \frac{7}{25}$ **d** $\frac{15}{50} = \frac{3}{10}$
e $\frac{8}{50} = \frac{4}{25}$ **f** $\frac{7}{50}$ **g** $\frac{9}{50}$ **h** $\frac{6}{50} = \frac{3}{25}$ **i** $\frac{6}{50} = \frac{3}{25}$

2 24, $\frac{1}{4}$ of 32 is 8 so there are 8 black counters

3 **a** 0.25 **b** 0.2 **c** 0.4 **d** 0.1 **e** 0.05 **f** 0.9

4 **a i** $\frac{1}{5}$ **ii** $\frac{2}{5}$ **b i** $\frac{1}{4}$ **ii** $\frac{1}{2}$

5 **a** $\frac{1}{3}$ **b** 3

6 **a** $\frac{1}{20}$ **b** green

7 Not A→0.8, not B→0.7, not C→0.4, not D→0

8 A = $\frac{3}{4}$ B = $\frac{2}{3}$ C = $\frac{1}{4}$ D = $\frac{9}{10}$ E = $\frac{13}{15}$ F = $\frac{1}{8}$ G = $\frac{2}{9}$

9 **a** $\frac{26}{52} = \frac{1}{2}$ **b** $\frac{4}{52} = \frac{1}{13}$ **c** $\frac{48}{52} = \frac{12}{13}$ **d** $\frac{13}{52} = \frac{1}{4}$ **e** $\frac{39}{52} = \frac{1}{4}$
f $\frac{48}{52} = \frac{12}{13}$ **g** $\frac{40}{52} = \frac{10}{13}$ **h** $\frac{48}{52} = \frac{12}{13}$ **i** $\frac{26}{52} = \frac{1}{2}$ **j** $\frac{32}{52} = \frac{8}{13}$ **k** $\frac{51}{52}$

10 **a** $\frac{10}{40} = \frac{1}{4}$ **b** $\frac{24}{40} = \frac{3}{5}$ **c** $\frac{30}{40} = \frac{3}{4}$ **d** 0
e $\frac{30}{40} = \frac{3}{4}$ **f** $\frac{16}{40} = \frac{2}{5}$ **g** $\frac{10}{40} = \frac{1}{4}$ **h** 1

Lesson 3.2 Mutually exclusive outcomes

Learning objective

- To recognise mutually exclusive events

Resources and homework

- Pupil Book 2.2, pages 55–59
- Homework Book 2, section 3.2
- Online homework 3.2, questions 1–10

Links to other subjects

- **History** –to know that certain events could not have happened at the same time

Key words

- intersection
- set
- Venn Diagram
- mutually exclusive
- union

Problem solving and reasoning help

- The **MR** questions 5, 6 and 7 in Exercise 3B of the Pupil Book link probability to financial skills. Encourage pupils to set out their tables of figures in a systematic way.

Common misconceptions and remediation

- Make certain that pupils understand that mutually exclusive events cannot happen at the same time; this will avoid confusion in later years with independent events.
- Pupils may also be confused by sentences with the words 'or' and 'and'. Carefully explain the meanings.

Probing questions

- Are Bruce Wayne and Batman mutually exclusive? (No)
- Give me examples of mutually exclusive events.
- How do you go about identifying all the mutually exclusive outcomes for an experiment?

Part 1

- To show the class outcomes that overlap, use a target board, as shown. Ask the class to sort the values into groups. Pupils will probably go for integers, fractions and decimals.
- Try to direct the class to ensure that the groups selected do not overlap.
- Now ask pupils to take, for example, the group of integers and break them into two sub-groups such as odd and even.

8	5	$\frac{1}{2}$	0.4
$\frac{3}{4}$	7	0.2	$\frac{1}{5}$
0.25	9	1	$\frac{2}{3}$

- Ask pupils to do this with the fractions, for example, numerators of 1 and numerators other than 1. Then ask them to do this with the decimals 0.2 with 0.4 (one decimal place) and 0.25 (two decimal places). Insist that no value must occur in more than one group (no overlap).
- Now say to the class that you will give them two groups from the target board and ask which values overlap. Give then, for example, prime and odd or the third column and second row.
- Ask the class to name two groups for which no values overlap.

Part 2

- Ask the class how many results there are when tossing a coin.
- Now ask pupils to write the combinations in their books.
- Question the class about the probability of each outcome.
- Explain that you cannot get two outcomes at the same time such as a head and a tail.
- Introduce the phrase 'mutually exclusive' as referring to outcomes that do not overlap.

- Ask pupils to help you rewrite an exhaustive list using only mutually exclusive outcomes. The challenge question at the end of Exercise 3B in the Pupil Book is a good example.
- Give some other examples of mutually exclusive events, but try to avoid using all events that are exact opposites of each other (for example, use rain, snow or fine, rather than rain and not rain). This is to avoid confusion with exhaustive events.
- The class could write down some examples in their books.
- **Pupils can now do Exercise 3B from Pupil Book 2.2.**

Part 3

- Using a pack of cards, ask pupils if the following pairs of events are mutually exclusive:
    - a queen and a club
    - an ace and a six
    - a king and a diamond
    - the ace of spades and a heart?

Answers

Exercise 3B

1 **a** Check pupils' diagrams
b Not exclusive due to 3 and 5 being in the intersection **c** $\frac{2}{6} = \frac{1}{3}$

2 **a** Check pupils' diagrams
b Not exclusive due to 2 and 3 being in the intersection **c** $\frac{2}{10} = \frac{1}{5}$

3 **a** Yes **b** No **c** No **d** No **e** No **f** No
g Yes **h** No **i** No **j** Yes **k** No **l** No

4 **a** No, any other colour may simply have not appeared in the sample **b** i, iii

5 **a** 1p, 2p, 3p, 5p, 6p, 7p, 8p **b** More because $\frac{4}{7}$ chance
c 1p, 2p, 3p, 5p, 6p, 7p, 8p, 10p, 11p, 12p, 13p, 15p, 16p, 17p, 18p
d Same chance as each other of $\frac{7}{15}$

6 **a** 3p, 6p, 7p, 11p, 12p, 15p, 21p, 22p, 25p, 30p, 51p, 52p, 55p, 60p, 70p, £1.01, £1.02, £1.05, £1.10, £1.20, £1.50, £2.01, £2.02, £2.05, £2.10, £2.20, £2.50, £3.00
b More than 60p, chance of $\frac{14}{28} = \frac{1}{2}$

7 **a**

+2	+1	3
+4	0	4
+4	−1	3
+4	+1	5
−3	0	−3
−3	−1	−4
−3	+1	−2

b **i** Positive ($\frac{6}{9}$) **ii** Odd ($\frac{5}{9}$)

Challenge: Four men run a race
Check pupils' answers.

Lesson 3.3 Using a sample space to calculate probabilities

Learning objective

- To use sample spaces to calculate probabilities

Resources and homework

- Pupil Book 2.2, pages 59–62
- Homework Book 2, section 3.3
- Online homework 3.3, questions 1–10

Links to other subjects

- **Science** –to record experimental data systematically in order to make predictions

Key word

- sample space

Problem solving and reasoning help

- The **PS** questions 7 and 8 of Exercise 3C in the Pupil Book require pupils to design their own sample space diagrams. **Less able** pupils may need help before beginning to work on these questions. Ask **more able** pupils to show the results in different sample space diagrams.

Common misconceptions and remediation

- Remind pupils that they will need to be very methodical when recording outcomes, otherwise they could miss some.

Probing questions

- What strategies do you use to make sure that you have found all the possible mutually exclusive outcomes for two successive events? For example: rolling two dice.
- How do you know that you have recorded all the possible outcomes?

Part 1

- Write a number on the board, for example, 36. Ask pupils to give you two numbers that add up to 36, two numbers with a difference of 36, two numbers with a product of 36, and finally, two numbers that divide to give an answer of 36.
- Now change the number and repeat.
- Next, take a sheet of A4 paper and tell pupils that you are going to fold it in various ways. Ask pupils to work out: Into how many pieces do the folds divide the sheet of paper?
- First, fold the sheet in half horizontally and vertically. Ask how many pieces there are.
- Now fold it in half again to obtain eight sections. Ask pupils if it makes any difference to the answer if you fold horizontally rather than vertically.
- Start with a fresh sheet of paper and fold it in half vertically and then into thirds (so that there are six columns) and then make one horizontal fold. Ask how many sections there are. (You could use this later for the coin and dice example.)

Part 2

- Tell the class about the boy who asked for three flavours on his ice cream cone: vanilla, strawberry and chocolate. The scoops were placed on top of each other. Ask pupils to tell you the different orders that the boy could have had.
- Prompt pupils to name different ways of recording the information: words, abbreviations or diagrams. Ask pupils to tell you the advantages or disadvantages of each method.

- Give pupils a different problem, for example: A boy and a girl each choose an apple or an orange. Ask them to record the different combinations in their books.
- At this stage introduce sample spaces.
- You can now use the paper from Part 1 to show pupils a sample space for a coin and a dice. Ask pupils to copy the 12 sections and label them.

H, 1	H, 2	H, 3	H, 4	H, 5	H, 6
T, 1	T, 2	T, 3	T, 4	T, 5	T, 6

- When pupils have completed this, tell them that you are thinking of one of these sections. Introduce the idea of the probability of pupils choosing the same section.
- **Pupils can now do Exercise 3C from Pupil Book 2.2.**

Part 3

- Write an empty sample space on the board and ask the pupils to tell you how to fill it in for a five-sided spinner numbered 1 to 5 and a six-sided dice, adding the totals.
- Ask pupils to give a few probabilities.
- Tell pupils that they are expected to know all the different methods of recording outcomes for two events. Reinforce the fact that there is no right or wrong way, but that some methods are better for some questions than others.
- Say to pupils that now that they have a good understanding of how to record, next time they will be able to look at more complicated probability questions involving more than one event.
- Remind pupils that when making lists the order matters sometimes, but not always. For example, it does not matter in which order the six lottery balls are picked out if they are your numbers. Make sure pupils understand the need to be systematic when making long lists.

Answers

Exercise 3C

1 **a** Check pupils' drawings

b **i** $\frac{1}{12}$ **ii** $\frac{3}{12} = \frac{1}{4}$ **iii** $\frac{4}{12} = \frac{1}{3}$

2 **a** BB, BG, GB, GG **b** Because it's $\frac{1}{4}$

3 **a** AA, AB, AP, BB, BP, PP **b** More pears in the bag than bananas

4 **a**

Plain	Beans
Cheese	Plain
Cheese	Cheese
Cheese	Beans
Beans	Plain
Beans	Cheese
Beans	Beans

b **i** $\frac{1}{3}$ **ii** $\frac{1}{3}$ **iii** $\frac{1}{9}$ **iv** $\frac{1}{9}$ **v** $\frac{1}{9}$ **vi** $\frac{1}{3}$ **vii** $\frac{4}{9}$ **viii** $\frac{2}{3}$

5 **a** 7 **b** **i** $\frac{3}{36} = \frac{1}{12}$ **ii** $\frac{4}{36} = \frac{1}{9}$ **iii** 0 **iv** $\frac{1}{36}$ **v** $\frac{15}{36} = \frac{5}{12}$

vi $\frac{21}{36} = \frac{7}{12}$ **vii** $\frac{6}{36} = \frac{1}{6}$ **viii** $\frac{18}{36} = \frac{1}{2}$ **ix** $\frac{10}{36} = \frac{5}{18}$ **x** $\frac{26}{36} = \frac{13}{18}$

6 **a** Check pupils' drawings **b** 12 **c** $\frac{10}{36} = \frac{5}{18}$

7 $\frac{3}{16}$

8 **a** Check pupils' diagrams **b** $\frac{10}{100} = \frac{1}{10}$

Problem solving: Odd socks

$\frac{1}{5}$

Lesson 3.4 Experimental probability

Learning objective
- To calculate probabilities from experiments

Resources and homework
- Pupil Book 2.2, pages 62–65
- Homework Book 2, section 3.4
- Online homework 3.4, questions 1–10

Links to other subjects
- **Science** –to make predictions based on experimental data
- **Geography** –to conduct social research

Key word
- experimental probability

Problem solving and reasoning help
- The **PS** questions 1 and 2 in Exercise 3D of the Pupil Book require pupils to have a good understanding of how experimental probability is used to solve a problem. They will need to analyse the questions and identify the mathematical methods needed to solve them. Pupils will also need to draw on their learning from the previous lessons in this chapter.

Common misconceptions and remediation
- Pupils often struggle to relate experimental data results to probabilities. Make sure pupils understand that experimental probabilities will be closer to the theoretical probability values if they increase the number of times they perform the experiment.

Probing questions
- What are the differences between using theoretical probability to find the probability of obtaining a 6 when rolling a dice, and performing an experiment for the same purpose?
- Which of the following statements are true or false?
  - Experimental probability is more reliable than theoretical probability.
  - Experimental probability gets closer to the theoretical probability, as more trials are carried out.

Part 1
- Use a multiplication table or write the first five rows on the board

1	**2**	**3**	**4**	**5**	**6**	**7**	**8**	**9**	**10**
2	4	6	8	10	12	*14*	16	18	20
3	6	9	12	15	18	*21*	24	27	30
4	8	12	16	20	24	28	32	36	40
5	10	15	20	25	30	35	40	45	50

- Ask a pupil to pick two numbers from the same column (for example, 14 and 21, in italics).
- Ask the class to say the numbers as a fraction (fourteen twenty-firsts). Then ask pupils to look at the beginning of the chosen rows and name the fraction they see (two-thirds).
- Ask pupils if they can tell you anything about these fractions. (They are equivalent.)
- Point out that this always works, for example $\frac{9}{36} = \frac{1}{4}$.
- Now pick out your own fractions and ask pupils to give you an equivalent fraction.

- You could then work this backwards by giving pupils a fraction in its simplest form and asking for an equivalent fraction. You could say, 'I want $\frac{3}{4}$ changed into sixteenths.'

Part 2

- Explain that you want to be able to predict whether the next person who comes through the door will have brown hair. Ask pupils to tell you how you could collect this information.
- On the board, write any key words that come up, for example, tally, observation sheet, survey, sample.
- Conduct a survey of the class by recording data about their hair colour. Use the results to obtain the experimental probability. Allow pupils time to write down this information.
- Ask them, in pairs or groups, to make a list of the types of information that they could collect to estimates probabilities. You may need to prompt pupils with a few examples such as: the number of times a six is thrown to test the fairness of a dice; the number of times the school bus is late this month; the sex of shoppers at a supermarket.
- **Pupils can now do Exercise 3D from Pupil Book 2.2.**

Part 3

- Ask what the difference is between 'theoretical probability' and 'experimental probability'.
- Remind pupils that experiments are prone to errors, but the more trials that are carried out, the more reliable the results tend to be.
- If appropriate, introduce the term 'relative frequency'.

Answers

Exercise 3D

1 **a** Greater chance to crash as it has a chance higher than $\frac{1}{2}$
b The 250 days as there are more results **c** 175 ÷ 250 **d** 0.7

2 **a** Yes, the 19 fives is much higher than you would expect
b Do more trials **c** 0.12 **d** 0.23 **e** 0.81

3 **a-c** Check pupils' charts **d** It should be getting closer to 0.5

4 **a** Check pupils' tables **b** $\frac{10}{20} = \frac{1}{2}$ **c** $\frac{4}{20} = \frac{1}{5}$
d Take it over a longer period of time

5 **a** 17 weeks and 6 days
b It is too small a sample
c $\frac{4}{5}$ **d** Yes, as 80% of the bulbs did last longer than 3000 hours

Problem solving: Roll the dice!

Answers will vary.

Review questions (Pupil Book pages 66–67)

- The review questions will help to determine pupils' abilities with regard to the material within Chapter 3. These questions also draw on the mathematics covered in earlier chapters of the book to encourage pupils to make links between different topics.
- The answers are on the next page of this Teacher Book.

Financial skills – Fun in the fairground (Pupil Book pages 68–69)

- Pupils extend their understanding of probability to a common real-life application that may be new to them. Pupils also make a real-life link between probability and financial skills.
- Look at the information in the Pupil Book. Write a fraction on the board, e.g. $\frac{1}{2}$.
- Ask pupils for other fractions that are the same as $\frac{1}{2}$, for example: $\frac{2}{4}$, $\frac{5}{10}$, $\frac{50}{100}$.

  Pupils should explain how they are getting the answers. Ask the class for something that has a probability of $\frac{1}{2}$, for example:
  - rolling a dice and getting a 'heads'
  - cutting a pack of cards for a red card
  - having a baby and it being a girl
  - rolling a dice to get an even number.
- Write $\frac{1}{5}$ on the board and ask for equivalent fractions, for example: $\frac{2}{10}$, $\frac{3}{15}$, $\frac{10}{50}$.
- Ask the class for something that has a probability of $\frac{1}{5}$, for example:
  - choosing one school day of the week at random
  - selecting a blue marble from a bag containing 5 marbles and only 1 is blue.
- Discuss fairgrounds and the different stalls with the class. Many stalls rely on the good chance that you will lose, but that someone will win a prize every now and again.

  This activity illustrates only two types of stalls. The data handling expected in this spread is:
  - reading from a table
  - probability
  - expectations.
- Before starting the activity put a table similar to this one on the board:

	Number of throws	Times hit target
Bull's eye	40	2
20	10	3
A double	50	5

- Say that someone was throwing darts at a dartboard for practice and kept a count of what they were aiming at and how many times they succeeded. Ask simple questions such as:
  - How many darts were thrown altogether? (100)
  - What is the probability of throwing a dart and hitting a: Bull's eye? ($\frac{2}{40} = \frac{1}{20}$); a 20? ($\frac{3}{10}$); a double? ($\frac{5}{50} = \frac{1}{10}$)
  - If 100 darts were thrown at the number 20, how many times would you expect them to have hit the target? ($100 \times \frac{3}{10} = 30$ times)
  - How many darts should be thrown at the Bull's eye to hit it once? ($\frac{40}{2} = 20$ times)
- Ask pupils to work through questions 1–10. Then ask if they feel that they have a chance of winning prizes in a fairground. The answer should be yes, but not the star prize many times.
- Ask pupils if they think that the Hook a Duck game seems fair.
- Encourage pupils' suggestions as to how they would change the game to make more profit.

Answers to Review questions

1 **a** 0.5 **b** 0.3 **c** 0.1 **d** 0.4

2 **a** triangles with angles 10,80,90 ; 20,90,70 ; 30,100,50 ; 40,110,30 ; 50,120,10

b **i** $\frac{2}{5}$ **ii** $\frac{2}{5}$ **c** $\frac{3}{5}$

3 $\frac{2}{5}$

4 **a** $\frac{2}{4}, \frac{2}{6}, \frac{2}{8}, \frac{2}{10}, \frac{2}{12}, \frac{4}{6}, \frac{4}{8}, \frac{4}{10}, \frac{4}{12}, \frac{6}{8}, \frac{6}{10}, \frac{6}{12}, \frac{8}{10}, \frac{8}{12}, \frac{10}{12}$

b $\frac{3}{15} = \frac{1}{5}$

5 **a** 10% **b** 0.2

6 **a** $\frac{2}{7}$ **b** $\frac{1}{7}$

7 For example –6, –5, –3, –1, 2, 4

8 **a** 2 **b** $\frac{5}{30} = \frac{1}{6}$

Answers to Financial skills – Fun in the fairground

1 £160

2 **a** $\frac{1}{40}$ **b** $\frac{1}{60}$ **c** $\frac{1}{320}$

3 **a** $\frac{7}{640}$ **b** $\frac{7}{128}$

4 **a** £6.25 **b** $\frac{5}{12}$

5 **a** 4 watches, 16 £10 notes, 8 £1 coins **b** £625 **c** £393

6 **a** $\frac{1}{45}$ **b** $\frac{1}{5}$ **c** $\frac{7}{9}$

7 $\frac{2}{9}$

8 **a** 5 **b** £2

9 **a** 250 **b** 6 **c** 50

10 **a** 625 **b** 486 **c** 125 **d** 14 **e** £137.94

4 Percentages

Learning objectives

- How to write one value as a percentage of another value
- How to use a multiplier to calculate a percentage increase or decrease
- How to write a change of value as a percentage increase or decrease
- How to use percentages to compare two quantities

Prior knowledge

- The equivalence between fractions, decimals and percentages
- How to calculate a percentage of a quantity, with or without a calculator

Context

- Percentage increase and decrease is probably one of the most common uses of mathematics in real life. Everyone meets it in some form or another even if this is only in terms of financial capability. To find real-life applications of percentages, look at the following financial education link: **http://www.pfeg.org/**
- Fractions, decimals and percentages are everywhere in real life and it is important for confidence and accuracy to be able to move between these different representations. This chapter reinforces the links between fractions, decimals and percentages.

Discussion points

- Which sets of equivalent fractions, decimals and percentages do you know?
- From one set that you know, for example, $\frac{1}{2}$ = 0.5 =50%, which others can you work out?
- How would you go about finding the percentage equivalents of any fraction?

Associated Collins ICT resources

- Chapter 4 interactive activities on Collins Connect online platform
- *Finding percentage increase and decrease* video on Collins Connect online platform

Curriculum references

Mathematical reasoning

- Interpret when the structure of a numerical problem requires additive, multiplicative or proportional reasoning

Solve problems

- Develop their mathematical knowledge, in part through solving problems and evaluating the outcomes, including multi-step problems
- Develop their use of formal mathematical knowledge to interpret and solve problems, including in financial mathematics

Number

- Define percentage as 'number of parts per hundred', interpret percentages and percentage changes as a fraction or a decimal, interpret these multiplicatively, express one quantity as a percentage of another, compare two quantities using percentages, and work with percentages greater than 100%
- Interpret fractions and percentages as operators

Fast-track for classes following a 2-year scheme of work

- Although pupils have met percentages before there are some important and quite challenging concepts in this chapter. The idea of percentages as a multiplier and the use multiplicative reasoning is very important to pupils' confidence and fluency with percentages. Therefore, while you may be able to leave out some of the earlier questions in each section, be careful about leaving out too much or moving on too fast.

Lesson 4.1 Calculating percentages

Learning objectives

- To write one quantity as a percentage of another
- To use percentages to compare quantities

Links to other subjects

- **Science** – to compare outcomes of experiments
- **Food technology** – to calculate nutritional components of pre-packed foods

Resources and homework

- Pupil Book 2.2, pages 71–73
- Intervention Book 2, pages 40–42
- Homework Book 2, section 4.1
- Online homework 4.1, questions 1–10

Key word

- percentage

Problem solving and reasoning help

- The challenge at the end of Exercise 4A in the Pupil Book requires pupils to use their understanding of percentages to make comparisons. Pupils also need to make links to representing data when presenting their results. You could model this challenge with simpler numbers. Pupils could present their findings to the class, who could provide constructive feedback based on an agreed set of criteria, for example, clarity of explanation, accuracy of results, and appropriate use of diagrams to enhance their explanation.

Common misconceptions and remediation

- Rather than simply learning a rule to calculate percentages of a quantity, pupils need to understand what they are doing and why. Pupils often confuse fractions and percentages, for example, by calculating 27% as one-twenty-seventh instead of 27 of 100. Make links to fractions and percentages so that pupils grasp the connection. Make sure pupils know what a percentage is and how to find a percentage of a quantity in a range of contexts.

Probing questions

- What fractions or percentages of given quantities can you work out easily in your head? Explain.
- When calculating percentages of quantities, from what percentages do you usually start? How do you use these percentages to work out other percentages?
- How do you decide when to use a calculator when finding percentages of quantities? Give me some examples of when you would and when you would not use a calculator.
- Explain how you use a calculator to find a percentage of a quantity or a fraction of a quantity.

Part 1

50%	30%	35%	40%	52%
55%	$17\frac{1}{2}$%	86%	$37\frac{1}{2}$%	70%
20%	25%	5%	48%	$12\frac{1}{2}$%
36%	18%	$77\frac{1}{2}$%	$87\frac{1}{2}$%	75%
$\frac{1}{2}$	$\frac{3}{10}$	$\frac{7}{20}$	$\frac{2}{5}$	$\frac{13}{25}$
$\frac{11}{20}$	$\frac{7}{40}$	$\frac{43}{50}$	$\frac{3}{8}$	$\frac{7}{10}$
$\frac{1}{5}$	$\frac{1}{4}$	$\frac{1}{20}$	$\frac{12}{25}$	$\frac{1}{8}$
$\frac{9}{25}$	$\frac{9}{50}$	$\frac{31}{40}$	$\frac{7}{8}$	$\frac{3}{4}$

- Say: '*Think of products that make 100*' (for example: 1 × 100, 2 × 50, 4 × 25, 5 × 20, 10 × 10).
- If possible, also establish $40 \times 2\frac{1}{2}$ and $8 \times 12\frac{1}{2}$.
- Use a target board, as shown, and ask pupils to give the percentage equivalent to each fraction.
- Discuss how to do this; by multiplying the denominator to make 100 and doing the same to the numerator.

Part 2

- This follows on from Part 1. Ask pupils for the percentage of 40 that is 32.
- Discuss ways of doing this, first without a calculator.
- Establish the fraction $\frac{32}{40}$ and multiply the top and bottom by $2\frac{1}{2}$ to give $\frac{80}{100}$, which is 80% (or cancel first to $\frac{4}{5}$). Also write on the board that 32 out of 40 is the same as 80 out of 100.
- Repeat the process with 55 out of 200 ($27\frac{1}{2}$%) and 3 out of 8 ($37\frac{1}{2}$%).
- Now discuss the calculator method to find, say, the percentage of 65 that is 13 (20%).
- This is the fraction $\frac{13}{65}$, which can be converted to a percentage by dividing through and multiplying by 100. Repeat with 24 out of 80 (30%); 55 out of 69 (80% when rounded off).
- Ask pupils how they would compare two sets of exam marks such as 38 out of 60, and 45 out of 70. Converting both to percentages gives 63.3% and 64.3%, so 45 out of 70 is a better percentage mark.
- **Pupils can now do Exercise 4A from Pupil Book 2.2.**

Part 3

- Write four quantities on the left-hand side of the board, for example: 45, 60, 56, 12.
- Write four quantities on the right-hand side of the board, for example: 120, 300, 200, 160.
- Match the quantities, one from the left-hand side and one from the right-hand side.
- Calculate, with or without a calculator, the percentage that the left value is of the right value.
- Some pairs are obvious, such as 12 and 120.
- Some are clearly non-calculator, such as 45 out of 300.
- Some have whole-number answers, but are not obvious, such as 56 out of 160 (35%).
- Discuss the appropriate methods to use for these.

Answers

Exercise 4A

1 **a** 50% **b** 80% **c** 90%
d 70% **e** 85% **f** 56%
g 95% **h** 76% **i** 38%

2 **a** 30% **b** 15% **c** 12%
d 10% **e** 6% **f** 3%

3 **a** 75% **b** 54% **c** 75%
d 40% **e** 65% **f** 45%

4 **a** 10% **b** 60% **c** 90%
d 70% **e** 20% **f** 5%

5 **a** 4% **b** 10% **c** 32% **d** 80% **e** 20%

6 **a** 40% **b** 40% **c** 4%
d 60% **e** 29% **f** 10%
Answers are given to the nearest whole number

7 **a** 46% **b** 25% **c** 92%
d 86% **e** 9% **f** 38%

8 **a** 58% **b** 58% **c** 68%
d 35% **e** 37% **f** 37%

9 8%

10 **a** 68%, 74% and 70%
b Maths has the highest percentage

11 **a** 48% **b** 87% **c** 9%

12 35%

13 25%

14 **a** 39% **b** 61%

15 **a** 43% **b** 57%

16 **a** 18% **b** 25%

17 **a** 78% **b** 21% **c** 1%

18 **a** Yes. 20% of 50 = 10
b No. There are 60 peopie in the room and 20% of 60 is 12, not 10

Challenge: What is in the waste?

A These are the percentages:

	Smith family	Jones family
Kitchen scraps	32%	29%
Plastics	8%	4%
Card or paper	4%	20%
Other	56%	47%

B Pie charts or a sectional bar chart are good choices

Lesson 4.2 Calculating percentage increases and decreases

Learning objective

- To use a multiplier to calculate a percentage change

Resources and homework

- Pupil Book 2.2, pages 73–76
- Homework Book 2, section 4.2
- Online homework 4.2, questions 1–10

Links to other subjects

- **Food technology** – to calculate dietary requirements for different people

Key word

- multiplier

Problem solving and reasoning help

- This lesson reinforces the concept of using percentage as an operator. This is an important step to ensure confidence and fluency, so make sure that you take time over it. It often helps to make links to fractions as operators. In the financial skills question at the end of Exercise 4B in the Pupil Book, pupils apply their understanding of percentage change to the familiar context of sales. This is a good opportunity to tackle any outstanding misconceptions.

Common misconceptions and remediation

- Pupils often are often confused when they come across percentages greater than 100. Using a real-life example here could help, starting with percentages with which they can work comfortably. Pupils need a good understanding of 100% as a whole before tackling percentage increases and decreases successfully.

Probing questions

- Explain how you would increase or decrease £18 e.g. by 25%. Can you do it in a different way? How would you find the multiplier for different percentage increases or decreases?

Part 1

- Use a target board, as shown.
- Recall the mental method for multiplying a two-digit number by a single-digit number (partitioning), for example: 7 × 23 = 7 × 20 + 7 × 3 = 140 + 21 = 161.
- Roll a dice (preferably a 10-sided dice). Ask pupils to multiply the numbers on the grid in turn by the number on the dice. Do this with individuals or ask the class to write down the answers (allow some jotting).

28	38	17	22	60
18	16	14	26	48
30	52	36	19	13
32	15	12	24	34

Part 2

- Ask pupils what they understand by percentage increases or decreases. Ask where such things occur in real-life (e.g., sales, increases from inflation, pay rises).
- Give an example such as: A cooker is reduced by 15% in a sale. If its original price was £320, what is its sale price? (You may need to introduce pupils to the vocabulary of cost, e.g. price, sale price, selling price.)
- Work through the example. First calculate 15% of £320 (= £48); then deduct this from the original price: £320 – £48 = £272.
- Now calculate 85% of £320 (= £272). Can pupils explain why this gives the same

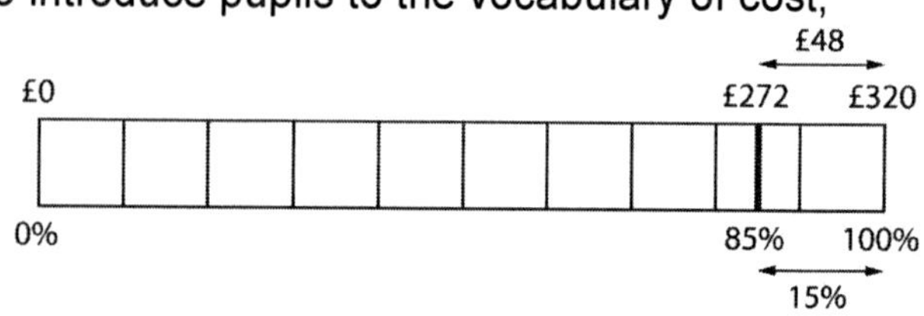

answer as deducting 15%? Show that decreasing a value by 15% is the same as finding 85% of the value (100 – 15 = 85), using a diagram like the one below.

- Give another example such as: A plumber adds VAT at 20% to her bills. How much will she add to a bill of £90? What will the final bill be? ($20\frac{1}{2}$% of £90 is £18, so the final bill will be £90 + £18 = £108)
- Now ask the class what the final bill is as a percentage of the original, that is £108 as a percentage of £90 (= 120%). Explain this by writing £90 + £18 = £108 on the board again. Ask what £90 is as a percentage of £90 (100%, write this underneath). Ask what £18 is as a percentage of £90 (20%, as just worked out).
- Give more examples if some pupils still do not grasp working with percentages over 100.
- Refer pupils to examples 3 and 4 on page 74 of the Pupil Book to reinforce the idea of percentages and decimals as operators.
- **Pupils can now do Exercise 4B from Pupil Book 2.2.**

Part 3

- Write a variety of percentages on the board, for example: 5%, 10%, 20%, 25%, and a variety of quantities such as: £32, 58 kg, 200 km and £150.
- Apply each percentage value to the different quantities and calculate a percentage increase or decrease, as appropriate. Most of these can be done without a calculator.

Answers

Exercise 4B

1	**a** 1.2	**b** £103.20			
2	**a** £38.40	**b** £73.20	**c** £220.80	**d** £11.28	
3	**a i** 1.3	**ii** 1.36	**iii** 1.43	**iv** 1.06	
	b i £93.60	**ii** £97.92	**iii** £102.96	**iv** £76.32	
4	**a** £48.30	**b** £56.70	**c** £77.70	**d** £81.90	
5	**a** £16.47	**b** £53.31	**c** £173.24	**d** £469.82	
6	**a** 49.98 kg	**b** 58.38 kg	**c** 66.78 kg	**d** 80.64 kg	
7	**a** 272.7 cm	**b** 353.7 cm	**c** 407.7 cm	**d** 461.7 cm	
8	**a** 0.85	**b i** £53.55	**ii** £44.63	**iii** £222.70	**iv** £50.99
9	**a** 0.9	**b** 0.7	**c** 0.63	**d** 0.57	**e** 0.25
10	**a** £69	**b** £23.40	**c** £17.70	**d** £45.48	
11	**a** £24.44	**b** 17.6 kg	**c** 38.08 minutes		
	d 418.5 ml	**e** 237.65 m	**f** 125 hours		

12 360 × 1.1 = 396 and 396 × 0.9 = 356.4 so the price is £356.40

13 **a** 54 720 **b** 59 280

Financial skills: Percentage reduction

A First row: 36, 28, 16; second row 67.50, 52.50, 30; third row 112.50, 87.50, 50; last row 315, 245, 140

B No, 60% off is better. People might make a mistake because 30 is less than 60 and confuse the amount off with the amount left after the reduction

Lesson 4.3 Calculating a change as a percentage

Learning objective

- To work out a change in value as a percentage increase or decrease

Resources and homework

- Pupil Book 2.2, pages 76–79
- Homework Book 2, section 4.3
- Online homework 4.3, questions 1–10

Links to other subjects

- **Food technology** – to calculate dietary requirement for different people
- **Geography** – to calculate comparative populations in different areas or over time

Key words

- No new key words for this topic

Problem solving and reasoning help

- This lesson continues to develop the concept of using percentage as an operator by looking at the inverse to calculate the percentage change or to calculate an initial value. This is an important concept. Pupils need to be fluent so that they are able to apply their understanding of percentages to real-life problems. Encourage discussion to challenge any misconceptions.

Common misconceptions and remediation

- Pupils are often confused when they come across percentages greater than 100. Pupils also make the mistake of pairing an increase with an equivalent decrease. For example, they think that a 50% increase followed by a decrease of 50% will take them back to their starting value. This misconception is using additive instead of multiplicative reasoning. Part 3 of this lesson tackles this misconception and pupils should draw on their work on inverse relationships during the lesson to explain their responses.

Probing questions

- The answer to a percentage increase question is £12. Make up an easy and a difficult question.

Part 1

- Revisit the outcomes of the last lesson by placing a new set of percentages and quantities on the board. As for Part 3 of the last lesson, ask pupils to apply each percentage value to the different quantities and calculate a percentage increase or decrease, as appropriate.
- Ask pupils to decide which ones can be done easily, without a calculator.
- Pupils could challenge each other to calculate easier percentages faster than their partner using a calculator.

Part 2

- Use examples 5 and 6 (page 77 of the Pupil Book) to demonstrate how to use multiplicative reasoning and inverse operations to calculate the percentage change between two values.
- **Pupils can now do Exercise 4C from Pupil Book 2.2.**

Part 3

- A shop decreases an item in a sale by 15%. When the sale is over they increase the same item by 15%. Is the price the same as it was before the sale? (No, the final multiplier is $\frac{99}{100}$ or 0.99.)

- Encourage pupils to use what they have done on percentages as a multiplier and the use of inverse relationships to explain their response to this question.
- Extension: What percentage increase would take the sale price back to its starting point ($\frac{10}{9}$).

Answers

Exercise 4C

1 **a** The multiplier is 378 ÷ 360 = 1.05 **b** 5%
2 **a** 1.6 **b** 60%
3 **a** 1.41 **b** 41%
4 65%
5 47%
6 7%
7 45%
8 **a** 0.97 **b** 3%
9 **a** 0.74 **b** 26%
10 **a** 62% **b** 34% **c** 17%
11 **a** 26% **b** 7% **c** 27%
12 5%
13 Red 5% increase, White 55% decrease, Blue 6% increase
14 **a** 8 cm **b** 12.8 cm
15 **a** £33 280 **b** £34 611.20

Problem solving: Five go on a diet

A 85 kg, 92 kg, 106 kg, 118 kg, 133 kg
B 85.5 kg, 92.15 kg, 105.45 kg, 116.85 kg, 131.1 kg
C Pupils' own answer, with reason. Answers may vary

Review questions (Pupil Book pages 80–81)

- The review questions will help to determine pupils' abilities with regard to the material within Chapter 4.
- These questions also draw on the mathematics covered in earlier chapters of the book to encourage pupils to make links between different topics.
- The answers are on the next page.

Challenge – Changes in population (Pupil Book pages 82–83)

- This activity is designed to give pupils the opportunity to demonstrate their understanding of percentage change to a real-life situation.
- All the information pupils need in order to answer the questions is provided, but they will need to read each question carefully to decide which information they need and what mathematical skills to use.
- If appropriate, you could model an example as part of guided group work.
- Ask pupils further questions as a warm-up to start working on questions 1–6 in the Pupil Book. Here are some examples:
  - What do you think the population of your town, UK, Europe, The world is?
  - How much do you think the population has changed since …?
  - What mathematical skills would you use to work out or calculate the change in population?
- You could give **more able** pupils more complex population data to study from websites such as 'Gapminder': **http://www.gapminder.org/** and 'Eurostat': **http://epp.eurostat.ec.europa.eu/portal/page/portal/eurostat/home/**

Answers to Review questions

1 83% (or 82.5%), 83% (or 82.7%) and 79% (or 78.9%)
2 55%
3 14% (or 13.7%)
4 £56.92
5 £56.70
6 81% and 73%
7 **a** 4% **b** 20%
8 **a** Add 21 **b i** 35% **ii** 26% **iii** 15% **iv** 13%
9 **a** 5% **b** 5% **c** 10% (or 10.25%)
10 **a** £267.50 **b** 1 + 7 ÷ 100 = 1.07
11 **a** 36% **b** 38%
12 **a** 61.5% **b** £75 400

Answers to Challenge – Changes in population

1 **a** 11.8% **b** 76.8% **c** From 2000 to 2010
d From 1970 to 1980 **e** Check pupils' diagrams
2 **a** 50.2 million **b** 66.0 million **c** 17.7%
3 **a** 18.5% **b** 15.4% **c** 36.7%
4 **a** 24.7% **b** 21.7%
5 **a** 12.9% **b** 11.4%
c Asia 23.4%; Africa 38.7%; Europe 9.3%; South America 19.0%; North America 15.8%
d Asia; No, Africa has the largest predicted percentage increase
6 The percentage of the people in the world who live in Europe has been decreasing since 1900, and is predicted to continue decreasing

5 Sequences

Learning objectives

- How to use flow diagrams to generate sequences
- How to use the *n*th term for sequences
- How to work out the *n*th term of a sequence
- How to use the special sequence of Fibonacci numbers

Prior knowledge

- How to create sequences and describe them, in words
- How to substitute numbers into an algebraic expression
- The sequence of square numbers
- The sequence of triangular numbers

Context

- The ability to generalise is crucial in a complex modern society. Being able to identify and generate number sequences is the first step towards progressing from the particular to the general in mathematics. Mathematics is all about the ability to see patterns, to hypothesise about these patterns and then seek to prove the hypothesis from first principles. Search on the internet for 'maths particular to the general' or look at: **http://nrich.maths.org/7831**. The article explores how to move from the particular to the general with KS3 pupils.

Discussion points

- Generate the first five terms:
    - start with 50 and subtract 5 each time
    - start with 3 and double each time.
- Use a spreadsheet to generate sequences and to explore the term-to-term and position-to-term rules.

Associated Collins ICT resources

- Chapter 5 interactive activities on Collins Connect online platform
- *Fibonacci poems* Wonder of Maths on Collins Connect online platform

Curriculum references

Develop fluency

- Use algebra to generalise the structure of arithmetic, including to formulate mathematical relationships

Reason mathematically

- Make and test conjectures about patterns and relationships; look for proofs or counter-examples

Algebra

- Generate terms of a sequence from either a term-to-term or a position-to-term rule
- Recognise arithmetic sequences and find the *n*th term
- Recognise geometric sequences and appreciate other sequences that arise

Fast-track for classes following a 2-year scheme of work

- Pupils can leave out Exercise 5A in the Pupil Book if they are familiar with the concept of flow diagrams. Pupils can also jump to the investigation on the *n*th term at the end of Exercise 5B if they have met this in Year 7.

Lesson 5.1 Using flow diagrams to generate sequences

Learning objective

- To use flow diagrams to generate sequences

Resources and homework

- Pupil Book 2.2, pages 85–87
- Intervention Workbook 1, pages 33–34
- Homework Book 2, section 5.1
- Online homework 5.1, questions 1–10

Links to other subjects

- **ICT** – to programme or code

Key words

- finite
- generate
- flow diagram
- infinite

Problem solving and reasoning help

- Flow diagrams give very precise instructions on how to perform a task or a calculation. In **PS** question 9 of Exercise 5A in the Pupil Book, pupils are required to generate sequences from patterns of dots. Ask pupils if they recognise any sequences. (Pupils should recognise square numbers and triangular numbers.)

Common misconceptions and remediation

- Pupils sometimes get confused with the shapes of boxes. Clearly explain the importance of each box and its use in a flow diagram.

Probing questions

- I am thinking of a sequence. It has these numbers in it: 12, 17, 22, 27.
  - What numbers come next in my sequence?
  - What numbers come before?
  - What clues did you use to work this out?
  - Give me a number that is greater than 40, which is in the sequence.
  - Tell me how you know that this number is in my sequence.
  - How could you check? Is 62 in the sequence?
  - How do you know?

Part 1

- This activity can be played as a game between teams.
- Ask a pupil to give the first two terms of a sequence, for example: 1, 5.
- Ask another pupil (from the other team if a game is played) to continue the sequence, for example: 1, 5, 9. Then ask another pupil (from the first team) to continue the sequence, for example:1, 5, 9, 13, …
- Once it becomes obvious that there is a well-defined sequence, stop and ask another pupil to give the first two terms of a sequence and so on. (Alternate the starting team.)
- Points are scored when:
  - Once you decide that a sequence is obvious, a pupil from the team that started can describe the rule.
  - At any time, pupils from the opposing team may challenge the last pupil to justify the term they have given and score a point if it is wrong or the rule is incorrect.
  - A pupil is unable to carry on the sequence (other team scores).

- Do not allow bizarre rules, but do encourage pupils to make sequences hard to spot. For example, 1, 5, 25, (× by 5), or 1, 2, 3, 5, 8, (Fibonacci), or 98, 97, 95, 92, (subtract 1, 2, 3,).
- If a score is kept, do not declare a winning team yet.

Part 2

- Write some sequences on the board such as:
  2, 5, 8, 11, 14,
  4, 8, 16, 32, 64,
  100, 99, 97, 94, 90,
- Say: 'Describe how these are building up.' 'What is the rule?' 'What are the next two terms?'
- Ask pupils to make up their own number sequences. (If number cards are available, pupils can lay out their patterns.) Choose some sequences to write on the board.
- When you have at least five sequences, ask the rest of the class to say what the next two terms in each sequence are, and also to describe the rule for developing the sequence. Introduce the idea of a term-to-term definition.
- Draw this flow diagram on the board:

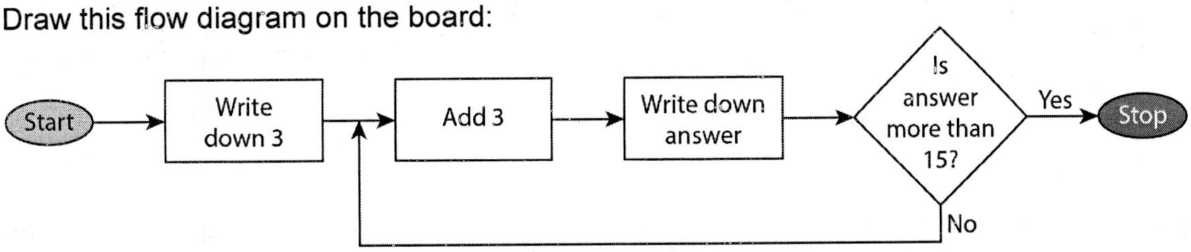

- Clearly explain the parts of the diagram [start/stop boxes (oval), action boxes (rectangular), decision boxes (diamonds)].
- Work through the diagram to give: 3, 6, 9, 12, 15, 18; the multiples of 3 up to 18.
- **Pupils can now do Exercise 5A from Pupil Book 2.2.**

Part 3

- Ask pupils to draw a flow diagram to generate:
  - the first 10 square numbers
  - the first 10 triangular numbers
- You could link this to question 9 of Exercise 5A.

Answers

Exercise 5A

1 **a** 3, 8, 13, 18, 23, 28, 33, 38, 43
b 10, 100, 1000, 10 000, 100 000, 1 000 000, 10 000 000
c 1, 4, 9, 16, 25, 36, 49, 64, 81, 100, 121
d 1, 3, 6, 10, 15, 21, 28, 36

2 Powers of 10

3 Square numbers

4 Triangle numbers

5 **a** Add 3
b Multiply by 4
c Increases by 3, 4, 5, 6, 7, ... or add consecutive whole numbers starting at 3
d Increases by 3, 5, 7, 9, ... or add consecutive odd numbers starting at 3

6 For example 1, 5, 9, 13, 17 (add 4); 1, 5, 25, 125, 625 (multiply by 5); 1, 5, 11, 19, 29 (add consecutive even numbers starting at 4)

7 **a** Increases 1, 2, 3, 4, 5, ... 61, 68
b Goes down 1, 2, 3, 4, ... 69, 62
c Increases 2, 4, 6, 8, 10, ... 43, 57
d Increases 4, 6, 8, 10, 12, ... 56, 72

8 **a** 1, 3, 9, 27, 81, 243

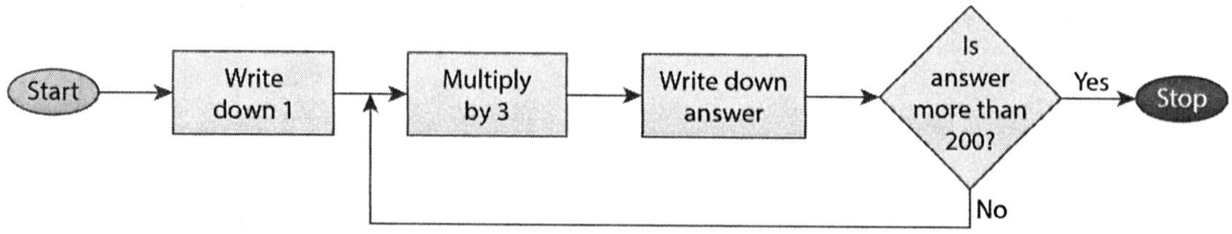

b 2, 4, 8, 16, 32, 64

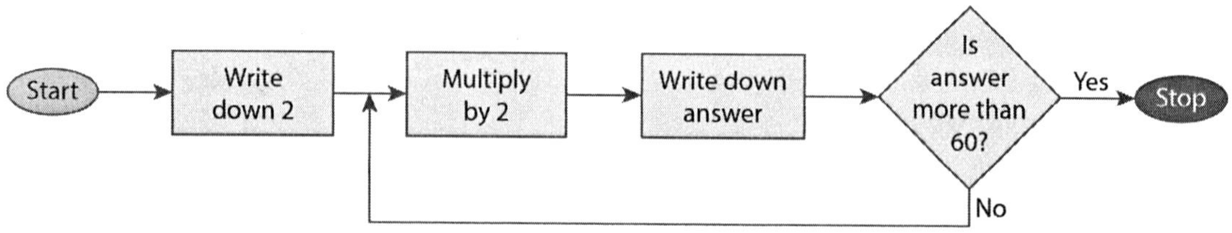

c 16, 8, 4, 2, 1, ½

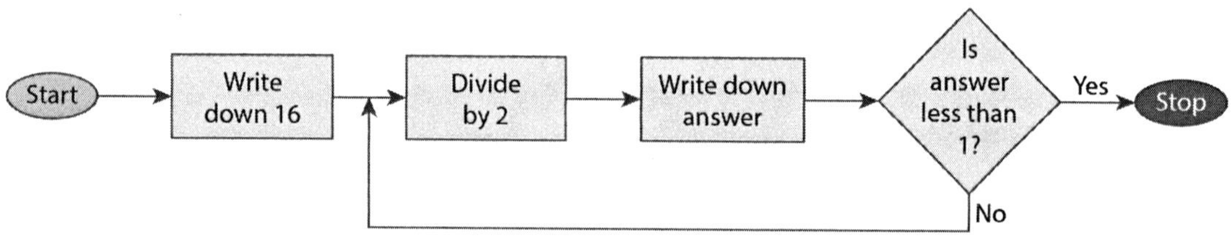

d 4, 1, −2, −5, −8, −11

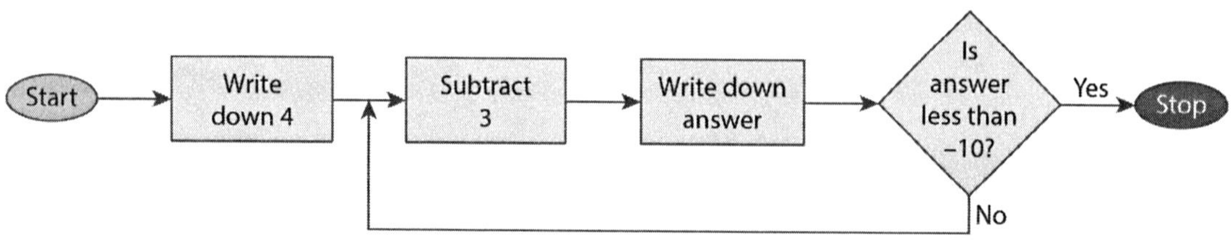

9 **a** 25, 36, 49, 64 **b** 15, 21, 28, 36 **c** 16, 32, 64, 128 **d** 30, 42, 56, 72

Problem solving: Algebraic flow diagrams

A

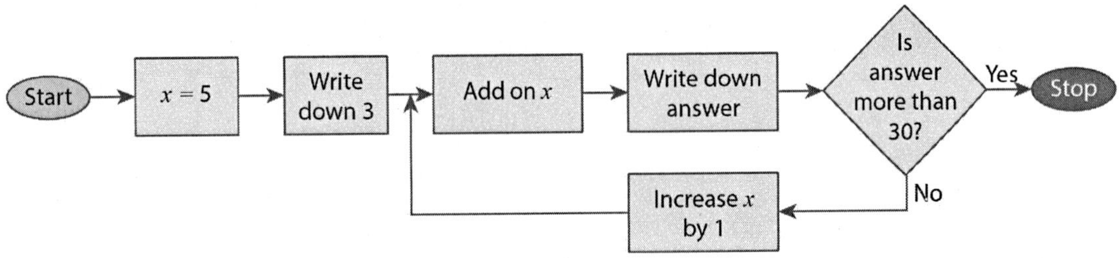

B

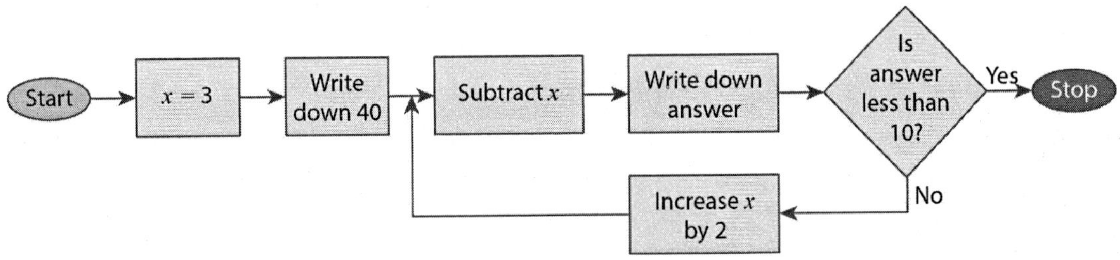

C

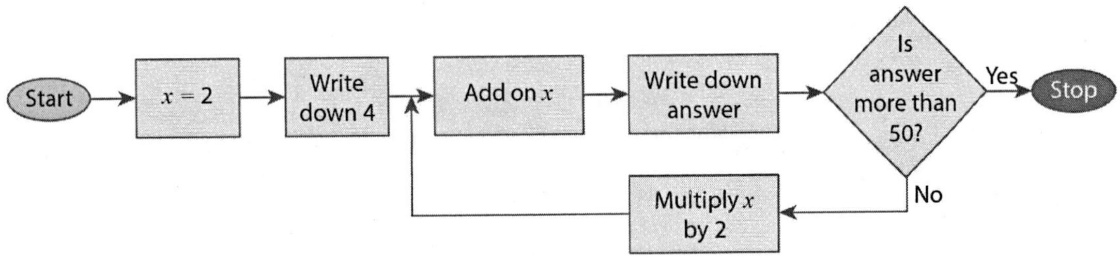

Lesson 5.2 The *n*th term of a sequence

Learning objective

- To use the *n*th term of a sequence

Resources and homework

- Pupil Book 2.2, pages 87–90
- Homework Book 2, section 5.2
- Online homework 5.2, questions 1–10

Links to other subjects

- **Science** –to use the *n*th term to work out an algebraic rule from the outcome of experiments

Key words

- algebraic expression
- coefficient
- constant term
- *n*th term

Problem solving and reasoning help

- **MR** question 5 of Exercise 5B in the Pupil Book requires pupils to investigate the link between the difference between terms and the coefficient of *n*. At this point, for **more able** pupils, it may be worth starting to explain how to write down the *n*th term, as this information is required to complete **PS** question 8.

Common misconceptions and remediation

- The most common error that pupils make when using the *n*th term rule is incorrectly substituting the term position number into the rule. Ensure that pupils are clear about how to substitute numbers into an expression. Pupils may also fail to understand that *n* will change for each new value of the sequence.

Probing questions

- The term-to-term rule for a sequence is +3. What does this tell you about the position-to-term rule?
- What do you look for in a sequence to help you find the position-to-term rule?

Part 1

- Write 5*n* on the board. Ask the class what this is. After taking some suggestions, focus on the idea that this is mathematical shorthand for 'five times the number that *n* represents'. Tell the class that this shorthand was invented by the earliest mathematicians to save time and space. Demonstrate that the expression has different values for different values of *n*. E.g., what is 5*n* worth when *n* is: 2, 3, 4, 9, 11, …?
- Repeat with 3*n*, 10*n*, …, until pupils are happy with the idea.
- Someone may suggest using a different letter. This is good! Encourage the idea that we can use any letter, but that some letters seem to be more useful than others. Ask why this might be. You could discuss the merits and disadvantages of the letters *n*, *x*, *p*, *o*, *l* and *m*.

Part 2

- Given a rule, can you generate the sequence? For example, the first term is 5 and the rule is: add 3 to the previous term. What is the sequence? (5, 8, 11, 14, 17, …)
- Other examples: first term 1, multiply each term by 5 (1, 5, 25, 125, 625, …); first term 2, multiply by –3 (2, –6, 18, –54, 162, …).
- Discuss the difference between this type of term-to-term definition and the type of rule used in Part 1 (a position-to-term definition). For example, in the first sequence above, any term is given by multiplying the term number by 3 and adding 2.

- Demonstrate that this rule works: The first term is 3 × 1 + 2 = 5, the second is 3 × 2 + 2 = 8, and the third term is 3 × 3 + 2 = 11.
- How can we write down this rule? E.g., 3 × term number + 2; this can be shortened to $3n$ + 2. This is the algebraic definition, which gives the nth term of the sequence.
- Example: each term of a sequence is given by $2n - 1$. What are the first five terms?
  2 × 1 – 1 = 1, 2 × 2 – 1 = 3, 2 × 3 – 1 = 5, 2 × 4 –1 = 7, 2 × 5 – 1 = 9
- E.g.: each term of a sequence is given by $4n + 4$. What are the first five terms now?
  4 × 1 + 4 = 8, 4 × 2 + 4 = 12, 4 × 3 + 4 = 16, 4 × 4 + 4 = 20, 4 × 5 + 4 = 24
- Point out that the sequences have a *constant difference* between terms, called *arithmetic sequences*. The expression for their nth term is always of this general form ($an \pm b$). How can we use the rule to find out how many slabs are around a 100 × 100 m^2 pond?
- **Pupils can now do Exercise 5B from Pupil Book 2.2.**

Part 3

- Ask for the meaning of 'generalisation'. How does the nth term demonstrate the nature of a pattern? Why is it more useful than the term-to-term rule?
- Ask what these generalisations generate:
  $2n$ (even numbers) — $n + 1$ (consecutive integers from 2)
  $2n - 1$ (odd numbers) — $6n$ (multiples of 6)
- Try some non-linear generalisations such as n^2 or $n(n + 1)$.

Answers

Exercise 5B

1 **a i** 7, 9, 11 **ii** 205 **b i** 3, 7, 11 **ii** 399 **c i** 2, 7, 12 **ii** 497
d i 5, 8, 11 **ii** 302 **e i** 9, 13,17 **ii** 405 **f i** 11, 21, 31 **ii** 1001
g i $2\frac{1}{2}$, 3, $3\frac{1}{2}$ **ii** 52 **h i** 6, 13, 20 **ii** 699 **i i** $\frac{1}{4}$, $\frac{3}{4}$, $1\frac{1}{4}$ **ii** $49\frac{3}{4}$
j i 9, 8, 7 **ii** –90 **k i** 18, 16, 14 **ii** –180 **l i** 4, 1, –2 **ii** –293

2 **a i** 3, 5, 7, 9 **ii** 3 **iii** 2 **b i** 4, 6, 8, 10 **ii** 4 **iii** 2
c i 5, 7, 9, 11 **ii** 5 **iii** 2 **d i** 6, 8, 10, 12 **ii** 6 **iii** 2

3 **a i** 4, 9, 14, 19 **ii** 4 **iii** 5 **b i** 7, 12, 17, 22 **ii** 7 **iii** 5
c i 1, 6, 11, 16 **ii** 1 **iii** 5 **d i** 8, 13, 18, 23 **ii** 8 **iii** 5

4 **a i** 2, 5, 8, 11 **ii** 3 **b i** 6, 10, 14, 18 **ii** 4
c i 2, 8, 14, 18 **ii** 6 **d i** 13, 23, 33, 43 **ii** 10

5 They are the same.

6 **a** $a = 4$, $d = 5$ **b** $a = 1$, $d = 2$ **c** $a = 3$, $d = 6$ **d** $a = 5$, $d = -2$

7 **a** 1, 8, 15, 22, 29, 36, … **b** 3, 5, 7, 9, 11, 13, … **c** 5, 9, 13, 17, 21, 25, …
d 0.5, 2, 3.5, 5, 6.5, 8, … **e** 4, 1, –2, –5, –8, –11, … **f** 2, 1.5, 1, 0.5, 0, –0.5, …

8 **a** 22 **b** $5n + 2$ **c** 252

Investigation: An *n*th term investigation

A

Sequence	First term, *a*	Difference, *d*	*n*th term	Coefficient of *n*	Constant term, *c*
5, 7, 9, 11, 13, …	5	2	$2n + 3$	2	3
8, 11, 14, 17, 20, …	8	3	$3n + 5$	3	5
8, 12, 16, 20, 24, …	8	4	$4n + 4$	4	4
7, 13, 19, 25, 31, …	7	6	$6n + 1$	6	1
4, 9, 14, 19, 24, …	4	5	$5n - 1$	5	-1
2, 6, 10, 14, 18, …	2	4	$4n - 2$	4	-2

B d = coefficient of n

C $c = a - d$

Lesson 5.3 Working out the *n*th term of a sequence

Learning objective

- To work out the *n*th term of a sequence

Resources and homework

- Pupil Book 2.2, pages 90–93
- Homework Book 2, section 5.3
- Online homework 5.3, questions 1–10

Links to other subjects

- **Science** –to use the *n*th term to work out an algebraic rule from the outcome of experiments

Key words

No new key words for this topic

Problem solving and reasoning help

- **PS** question 3 produces odd and even number sequences. Pupils will need to work out the *n*th term rule for both sequences in order to answer parts b and c, and then add the rules together in order to answer part d. **MR** question 8 requires pupils to complete a flow diagram in order to generate several sequences. Pupils will need a good understanding of how to produce the *n*th term rule from a sequence in order to complete this question.

Common misconceptions and remediation

- The most common error in finding the *n*th term of an arithmetic sequence is simply to write n plus the difference. For example, for the sequence 3, 7, 11, 15 … the error is to write $n + 4$.

Probing questions

- The term-to-term rule for a sequence is –3. What does this tell you about the *n*th term?
- What do you look for in a sequence to help you find the *n*th term?
- How would you find the *n*th term rule for the information relating to the following sequence?
  - Position 3 5 10
  - Term 1 19 39

Part 1

- Ask pupils to make up a number, for example, 25.To do this, pupils should say a number to which you apply a secret rule; then you say the number. For example, if the rule is 'add 2', pupils may say 7; you reply 9; they say 16; you reply 18. Eventually a pupil will say 23 to which you reply 25.
- This is a fairly easy rule. Other rules (to make you say '25') are 'square, 5', '$3n + 1$, 8', 'multiply by 5, 5' and so on.
- To tie in with Part 2, the rules should be of the type $ax \pm b$.

Part 2

- On the board, write the sequence: 3, 7, 11, 15, 19, ...
- Ask the class if they can name the *n*th term of this sequence. It is unlikely, but someone may spot that it is $4n - 1$. If not, write this on the board as the *n*th term.
- Explain that $4n - 1$ is known as the general term or *n*th term of the sequence.
- Ask the class if they know the *n*th term of 7, 12, 17, 22, 27, ... (Establish that it is $5n + 2$)
- Leave the *n*th terms and the sequences visible and ask the class to see if they can spot any connections.
- Now put another sequence on the board for example 2, 5, 8, 11, 14, ... Ask if anyone knows what the *n*th term is. If a pupil knows this ($3n - 1$), ask him or her to explain how they know.

- If not, or if the explanation does not cover this, make sure that the class can see that the '*n*' term is generated by the difference. Use the examples already seen to emphasise this.
- Also, make sure that the class can see that the constant term is the first term minus the difference. Again, use the examples already seen to emphasise this.
- Put other sequences on the board and ask pupils to use the rules to generate the *n*th terms.
- **Pupils can now do Exercise 5C from Pupil Book 2.2.**

Part 3

- Ask for the meaning of 'generalisation'. How does the *n*th term demonstrate the nature of a pattern?
- Why is it more powerful than the simple term-to-term rule?
- Ask what the generalisations are for:
  - even numbers; consecutive integers; odd numbers; multiples of 5.
- Write the square numbers on the board and ask for the generalisation of this pattern.

Answers

Exercise 5C

1 **a** $6n-2$ **b** $3n+6$ **c** $6n+3$ **d** $3n-1$ **e** $7n-5$
f $2n+6$ **g** $4n+6$ **h** $8n-5$ **i** $10n-1$ **j** $9n-5$

2 **a i** $4n+1$ **ii** 201 **b i** $5n+1$ **ii** 251 **c i** $3n-1$ **ii** 149 **d i** $7n+1$ **ii** 351

3 a

Pattern	Number of blue squares	Number of red squares
1	1	2
2	3	4
3	5	6
4	7	8

b $2n-1$ **c** $2n$ **d** $4n-1$

4 **a** $95-5n$ **b** $50-7n$ **c** $31-3n$ **d** $52-8n$

5 **a** $0.5n+2$ **b** $2.5n+8$ **c** $0.1n+3$ **d** $8.2-0.2n$

6 **a i** $4n$ **ii** $3n+1$ **iii** $7n+1$ **b i** 80 **ii** 61 **iii** 141

7 **a** £31 **b** £$(3n+1)$

8 These are examples. Other answers are possible:
a a= 2, add 2,l =10: 2, 4, 6, 8, 10
b a= 1, add 2,l = 9: 1, 3, 5, 7, 9
c a= 5, add 5,l = 25: 5, 10, 15, 20, 25
d a= 1, add 2, 3, 4, …,l = 15: 1, 3, 6, 10, 15
e a= 1, add 10, l= 51: 1, 11, 21, 31, 41, 51
f a= 1, add 3,l = 13: 1, 4, 7, 10, 13
g a= 1, multiply by –2, l= 16: 1, –2, 4, –8, 16

Mathematical reasoning: Square sequences

A n^2
B n^2+1
C n^2-1
D $\frac{1}{2}n^2$
E n^2+n

Lesson 5.4 The Fibonacci sequence

Learning objective

- To know and understand the Fibonacci sequence

Resources and homework

- Pupil Book 2.2, pages 94–95
- Homework Book 2, section 5.4
- Online homework 5.4, questions 1–10

Links to other subjects

- **Physical education** – to prepare for a tournament that uses a round robin grouping strategy; the number of matches played between *n* teams will be the *n*–1 triangular number
- **Design and technology** – to use the golden ratio from the Fibonacci sequence for design purposes
- **Science** – to use Fibonacci patterns in nature (the most common number of flower petals is a Fibonacci number)

Key words

- Fibonacci sequence

Problem solving and reasoning help

- **PS** questions 6 and 7 of Exercise 5D in the Pupil Book require pupils to generate sequences and spot any patterns based on variations of the Fibonacci sequence. **MR** question 8 is designed to produce the 'golden ratio'. Explain the golden ratio to pupils and how it is found in architecture and the art and design world. Suggest that pupils do an internet search for 'golden ratio', 'Fibonacci', and each word with 'architecture' and 'art and design'.

Common misconceptions and remediation

- Pupils simply need to remember that the Fibonacci sequence is generated by summing the previous two terms. Pupils often struggle to start the sequence as the first term is 1 and there is no previous term. Explain to pupils that the term before the first 1 is 0 and 0+1=1, which gives the second term.

Probing questions

- How many different ways can you go down 12 steps taking one at a time or two at a time? You could refer to this link 'Step 2 Step' at: **http://nrich.maths.org/1019**.

Part 1

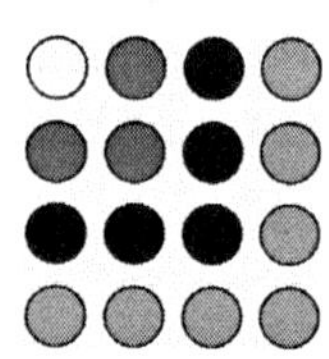

- The sum of consecutive odd numbers can be represented as square arrays. The picture shows an example.
- The sum of the first four consecutive numbers is: 1 + 3 + 5 + 7 = …
- Can you use the picture to work out the answer without adding all the numbers?
- Explain how you have done this (e.g., four across and four down = $4 \times 4 = 16$).
- Use grouping to support and/or challenge this activity for **less able** and **more able** pupils, respectively.

Part 2

- Introduce this topic using Fibonacci's original problem.

- Start with a pair of rabbits, (one male and one female), born on January 1. Assume that all months are of equal length and that:
    - o Rabbits begin to reproduce two months after they are born.
    - o At the age of two months, each pair of rabbits produces a mixed pair, (one male, one female), and then another mixed pair each month thereafter.
    - o No rabbits die.
- How many pairs of rabbits will there be after one year?
  Fibonacci's solution
- *1, 1, 2, 3, 5, 8, 13, 21, 34, 55, 89, 144, ...*
- At the end of the year, there will be 144 pairs of rabbits, all resulting from the original pair born on January 1 of that year.
- **Pupils can now do Exercise 5D from Pupil Book 2.2.**

Part 3

- Ask pupils what the sixth term of the Fibonacci-type sequence is, which starts with 2 and 38 as the first two terms.
- How many Fibonacci type sequences can they find containing the number 196 as one of the terms where the sequence starts with two whole numbers *a* and *b* with $a<b$?

Answers

Exercise 5D

1 1, 1, 2, 3, 5, 8, 13, 21, 34, 55, 89, 144, 233, 377, 610
2 1, 144
3 1, 3, 21, 55
4 2, 3, 5, 13, 89, 233
5 **a** 26, 42, 68, 110 **b** 39, 63, 102, 165 **c** 29, 47, 76, 123 **d** 42, 68, 110, 178
6 **a i** 4 **ii** 7 **iii** 12
b The answers are all 1 less than two terms further on in the sequence
7 e.g. Take 5, 8, 13 21. 5 × 21 = 105, 8 × 13 = 104. Difference = 1.
Take 21, 34, 55, 89. 21 × 89 = 1869. 34 × 55 = 1870. Difference = 1.
Take 2, 3, 5, 8. 2 × 8 = 16. 3 × 5 = 15. Difference = 1. The rule always works
8 **a**

Term	Previous term	Answer
1	1	1
2	1	2
3	2	1.5
5	3	1.66666...
8	5	1.6
13	8	1.625
21	13	1.61538...
34	21	1.61904...
55	34	1.61764...
89	55	1.61818...

b The numbers are getting closer to the decimal 1.618 .... This number is known as the 'Golden ratio' and is1.618 033988 749894 84820 to 20 decimal places

Investigation: Steps and stairs

A 3 (1+1+1, 1+2, 2+1)
B 5 (1+1+1+1, 1+1+2, 1+2+1, 2+1+1, 2+2)
C

Number of stairs	Number of ways
1	1
2	2
3	3
4	5
5	8

D The numbers are in the Fibonacci sequence
E There are 377 ways of going up 13 stairs. The sequence is 1, 2, 3, 5, 8, 13, 21, 34, 55, 89, 144, 233, 377

Review questions

(Pupil Book pages 96–97)

- The review questions will help to determine pupils' abilities with regard to the material within Chapter 5.
- These questions also draw on the mathematics covered in earlier chapters of the book to encourage pupils to make links between different topics.
- The answers are on the next page of this Teacher Guide.

Investigation – Pond borders

(Pupil Book pages 98–99)

- Pupils could tackle this investigation after completing Lesson 5.3.
- Tell the class that you have a rectangular pond that is $24m^2$.
- Now ask them what are the possible dimensions if the sides are whole numbers? (You should get 1×24, 2×12, 3×8, 4×6
- For steps 1 to 4, make sure pupils start with a pond of $1m^2$ and work up through the square numbers. Encourage pupils to draw an extended table and work out the term-to-term rule. Once they have a term-to-term rule, they can find the *n*th term.
- Make sure that pupils can see that the '*n*' term is generated by the difference. Also make sure that the class can see that the constant term is the first term minus the difference.
- Write other sequences on the board and ask pupils to use the rules to generate the *n*th terms.
- Encourage pupils to explain what they are doing at each stage of the investigation and to draw tables starting with the smallest pond. Pupils should then work out the *n*th term from the table following the method above.
- Encourage pupils to explain in words how they work out the number of tiles from studying the patterns. They might say:
    - pond length × 2 + pond width × 2 + 4 corners
    - perimeter + 4
- Ask pupils to come up with an expression to calculate the border of any rectangular pool
    - $2x + 2y + 4$
    - $2(x + y) + 4$

  Where x and y are the length and width respectively.

Answers to Review questions

1	**a** 3, 5, 7, 9, 11, 13, 15, 17, 19		**b** 10, 5, 0, –5, –10, –15, –20, –25		
	c 1, –2, 4, –8, 16, –32, 64, –128, 256		**d** 5000, 500, 50, 5, 0.5, 0.05		
2	**a** 3, 8, 13, 18, 23, 28		**b** 4, 6.5, 9, 11.5, 14, 16.5		
	c 10, 6, 2, −2, −6, −10		**d** −5, −8, −11, −14, −17, −20		
3	**a** £420	**b** £440	**c** £460	**d** £480	**e** $20n$ + £400
4	**a** $6n - 3$	**b** $3n + 7$	**c** $6n + 1$	**d** $3n - 1$	
5	**a** $5n + 1$	**b** 51	**c** 20	**d** 39, 4	
6	**a i** Add 1	**ii** $4n + 2$	**b i** Divide by 2	**ii** $3n + 3$	
	c i Subtract 4	**ii** $19 - 3n$	**d i** Multiply by2	**ii** $10n - 6$	
7	**a i** 2, 5, 10, 17	**ii** 401	**b i** 4, 9, 16, 25	**ii** 441	
	c i 4, 7, 12, 19	**ii** 403	**d i** 4, 10, 18, 28	**ii** 460	
	e i 4, 11, 30, 67	**ii** 8003			

Answers to Investigation – Pond borders

A	**a** 18	**b** $2n + 6$	**c** 46
B	**a** 22	**b** $2n + 8$	**c** 48

6 Area of 2D and 3D shapes

Learning objectives

- How to work out the areas of triangles, parallelograms and trapezia
- How to work out the areas of compound shapes
- How to work out the surface areas of cuboids

Prior knowledge

- How to work out the perimeters and areas of squares and rectangles
- How to work out the volumes of cubes and cuboids

Context

- Remind pupils that perimeter, area and volume are used widely in many jobs and professions, from farming to astronomy. Encourage pupils to ask family and friends if they use these units of measure in their work. Pupils could also explore specific jobs on the internet. A good example is the building industry, which is totally dependent on workers being able to measure lengths and calculate areas.

Discussion points

- In what way are the perimeter and area of a shape different?
- Do you have any tips to help you to remember which is which?

Associated Collins ICT resources

- Chapter 6 interactive activities on Collins Connect online platform
- *Feeling the cold* Wonder of Maths on Collins Connect online platform

Curriculum references

Geometry and measures

- Use the properties of faces, surfaces, edges and vertices of cubes, cuboids, prisms, cylinders, pyramids, cones and spheres to solve problems in 3D
- Derive and apply formulae to calculate and solve problems involving: perimeter and area of triangles, parallelograms, trapezia, volume of cuboids (including cubes) and other prisms (including cylinders)
- Calculate and solve problems involving: perimeter of 2D shapes (including circles), areas of circles and composite shapes

Develop fluency

- Use language and properties precisely to analyse 2D and 3D shapes

Solve problems

- Develop their mathematical knowledge, in part through solving problems and evaluating the outcomes, including multi-step problems

Fast-track for classes following a 2-year scheme of work

- Pupils should be familiar with many of the concepts in this chapter. Check their understanding with a couple of examples and then move on to the **MR**, **PS** and challenge or investigation questions in each lesson.
- You may also want to leave out Lesson 6.1 and combine Lesson 6.2 and Lesson 6.3.

Lesson 6.1 Area of a triangle

Learning objective

- To work out the area of a triangle

Resources and homework

- Pupil Book 2.2, pages 101–104
- Homework Book 2, section 6.1
- Online homework 6.1, questions 1–10

Links to other subjects

- **Design and technology** – to calculate areas of shapes as part of the design aspect of a project

Key words

- base
- height
- compound shape
- perpendicular height

Problem solving and reasoning help

- **PS** question 7 of Exercise 6A in the Pupil Book requires pupils to combine the new knowledge form this lesson with their existing knowledge to solve a multi-step problem. To help **less able** pupils, you may want to model the process with a similar example. Talk through the questions you would ask yourself to answer the question. **MR** question 8 and the investigation at the end of Exercise 6A give pupils the opportunity to make links to other areas of mathematics; in this case, number and sequences.

Common misconceptions and remediation

- Pupils should understand that the height of a triangle is the vertical height, not the length of a side (except in some specific examples). After mastering the area of a triangle, **less able** pupils may divide the area of every shape by 2; watch out for this.

Probing questions

- The area of a triangle is 18 cm^2. What are the possible lengths of the base and the height?
- Right-angled triangles have half the area of the rectangle with the same base and height. What about non-right-angled triangles?

Part 1

- This is a revision starter on perimeters and areas of rectangles. Ask pupils to form pairs. Each pair should have their own whiteboards or sheets of A4 paper.
- Ask pairs to sketch a rectangle that has a perimeter of 10 cm and an area of 6 cm^2, and then to hold up their whiteboards or sheets of paper.
- The first pair to give the correct answer gains a point.
- Repeat the activity a number of times using different examples to suit the ability of the class.

Part 2

- For this activity the class needs coloured card or gummed paper and scissors.
- Ask pupils to cut out two identical triangles.
- Cut both triangles into two, as in the diagram.
- Arrange the four triangles to form a rectangle, which shows that the area of the triangle is equal to half the area of the rectangle: Area 1 = Area 2 and Area 3 = Area 4
- We define the base and the height of a triangle as in the diagram below the rectangle.

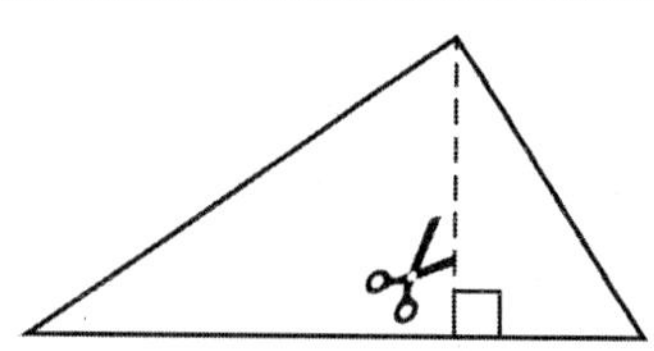

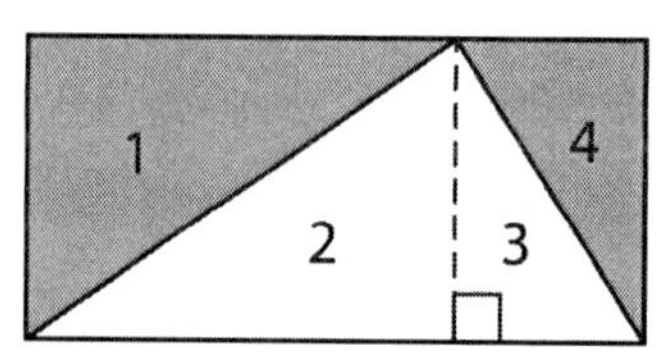

- The height of the triangle is sometimes referred to as the *perpendicular height.*
- The area of a triangle is given by the formula:

  $A = \frac{1}{2} \times b \times h = \frac{1}{2}bh$

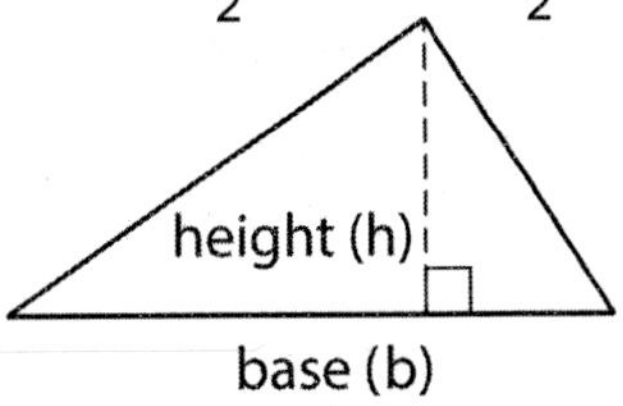

- Show the class an example.
- On the diagram, $b = 6$ cm and $h = 4$ cm. So $A = \frac{1}{2} \times 6 \times 4 = 12$ cm^2.

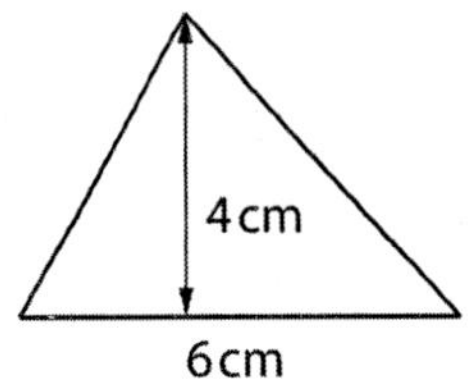

- Explain that this calculation can be worked out in different ways, for example: $(6 \div 2) \times 4 = 12$ or $(6 \times 4) \div 2 = 12$.
- Explain that the height of the triangle is sometimes given outside the triangle. For example, on the diagram to the right $b = 8$ cm and $h = 5$ cm, so $A = \frac{1}{2} \times 8 \times 5 = 20$ cm^2.

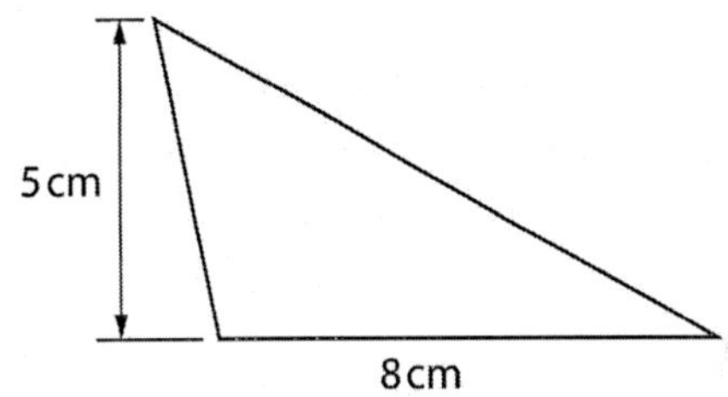

- Show the class how to find the area of a compound shape by dividing it into rectangles and triangles, as in the shape shown.
- Divide the shape into a rectangle and a triangle.

  Area of A = $\frac{3 \times 4}{2} = \frac{12}{2} = 6$ cm^2.

  Area of B = $3 \times 5 = 15$ cm^2.

  So the area of the shape = $6 + 15 = 21$ cm^2.
- **Pupils can now do Exercise 6A from Pupil Book 2.2.**

Part 3

- Ask the class to work in pairs or small groups.
- Ask each group to explain how to find the area of a triangle by drawing diagrams on individual whiteboards or on A4 paper.
- Ask different groups to show their explanation to the rest of the class.

Answers

Exercise 6A

1	**a** 35 cm^2	**b** 108 cm^2	**c** 24.5 cm^2	**d** 6m^2	**e** 28m^2
2	**a** 15 cm^2	**b** 60 cm^2	**c** 270mm^2		
3	**a** 17.5 cm^2	**b** 30m^2	**c** 120mm^2		
4	**a** 15cm^2	**b** 28cm^2	**c** 27.5m^2	**d** 8mm	**e** 7m
5	**a** 6cm^2	**b** 10 cm^2	**c** 6cm^2	**d** 12 cm^2	
6	**a** 6m^2	**b** 45 cm^2	**c** 12m^2		
7	72 m^2				

8 Triangles for which the product of the base and the height is 72
e.g., 1×72, 2×36, 3×24, 4×18, 6×12, 8×9

Investigation: Compound triangles

A $A = 1 + 3 + 6 + 10 = 20$cm^2

B The sequence is 1, 3, 6, 10 and the rule for the sequence is 'add on 1 more each time'. The next number in the sequence is 15

C The area of the new shape is $20 + 15 = 35$cm^2

Lesson 6.2 Area of a parallelogram

Learning objective

- To work out the area of a parallelogram

Resources and homework

- Pupil Book 2.2, pages 105–107
- Homework Book 2, section 6.2
- Online homework 6.2, questions 1–10

Links to other subjects

- **Design and technology** – to calculate areas of shapes as part of the design aspect of a project

Key word

- parallelogram

Problem solving and reasoning help

- **MR** question 9 of Exercise 6B in the Pupil Book and the challenge question at the end of the exercise give pupils the opportunity to make links from their new learning in this lesson, the previous lesson and other areas of understanding. Pupils then extend this to similar ideas, which they may not yet have encountered; in this case, working out the area of a rhombus. Be explicit about pupils' use of prior knowledge, to help them gain new understanding.

Common misconceptions and remediation

- Pupils should understand that the height of a parallelogram is the vertical height, not the length of a side.

Probing questions

- Why do we find the area of a parallelogram by multiplying the base by the perpendicular height?
- Can you convince me that a rhombus must be a parallelogram, but a parallelogram is not necessarily a rhombus?

Part 1

- On the board, draw the grid as shown or on an OHP.
- Ask the class to work out the largest number they can make by adding any two adjacent numbers on the grid. Explain that the two numbers can be from any row, column or diagonal.

1	8	5
4	2	7
9	6	3

- Repeat this, but change the operation to multiplication.
- Repeat the activity, but now use three adjacent numbers from any row, column or diagonal.
- You could extend this activity by using larger numbers or grids of different sizes.

Part 2

- For this activity, the class needs coloured card or gummed paper and scissors.
- Ask the class to cut out a parallelogram as shown.
- Cut off a triangle as in the diagram and place it at the other side. This shows that the area of the parallelogram has the same area as a rectangle with the same base and height.

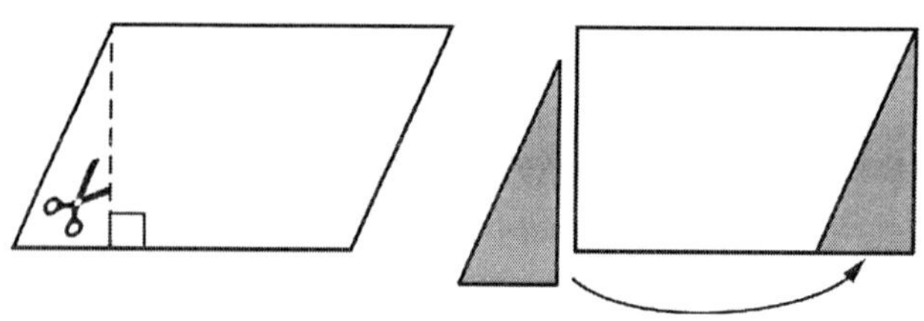

- The area of a parallelogram is given by the formula:
  $A = b \times h = bh$

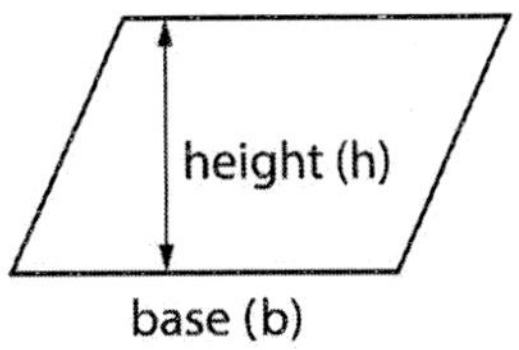

- Show the class an example. On the diagram, b = 8 cm and h = 5 cm, so $A = 8 \times 5 = 40$ cm^2.

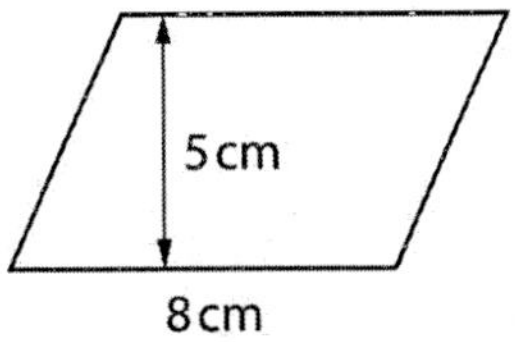

- **Pupils can now do Exercise 6B from Pupil Book 2.2.**

Part 3

- Ask the class to work in pairs or small groups. Ask each group to explain how to find the area of a parallelogram by drawing diagrams on individual whiteboards or on A4 paper.
- Ask different groups to show their explanations to the rest of the class.

Answers

Exercise 6B

1	**a** 36 cm^2	**b** 150 cm^2	**c** 768mm^2		
2	**a** 80 cm^2	**b** 49m^2	**c** 80 cm^2		
3	**a** 88.2 cm^2	**b** 30 cm^2			
4	**a** 32 cm^2	**b** 204 cm^2	**c** 40m^2	**d** 4mm	**e** 3.5m
5	**a** 20 cm^2	**b** 15 cm^2	**c** 24 cm^2		
6	4.5 cm				
7	6.25cm				
8	**a** 34cm	**b** 57cm^2			
9	6cm				

Challenge: Area of a rhombus

a	38.5 cm^2	**b** 96 cm^2	**c** 720mm^2

Lesson 6.3 Area of a trapezium

Learning objective
- To work out the area of a trapezium

Resources and homework
- Pupil Book 2.2, pages 108–111
- Homework Book 2, section 6.3
- Online homework 6.3, questions 1–10

Links to other subjects
- **Design and technology** – to calculate areas of shapes as part of the design aspect of a project

Key word
- trapezium

Problem solving and reasoning help
- **MR** question 9 and **PS** question 10 of Exercise 6C in the Pupil Book require pupils to make links across strands of mathematics to solve a multi-step, slightly more complex problem.

Common misconceptions and remediation
- Pupils should understand that the height of a trapezium is the vertical height, not the length of a side (except in some specific examples).

Probing questions
- What properties do you need to know about a quadrilateral in order to be sure that it is: a parallelogram; a rhombus; a trapezium; an isosceles trapezium?
- Why can a trapezium not have three acute angles?
- Which quadrilateral can have three acute angles?

Part 1
- Ask the class to work in pairs or groups for this activity.
- Explain to the class that they have to make as many numbers from 1 to 10 as they can in about 10 minutes by using only four 4s and any of the mathematical operations +, –, ×, ÷, which may be repeated. Allow them to use the rules of BIDMAS, but brackets may be used.
- Some examples are: $4 \div 4 \times 4 \div 4 = 1$; $4 \div 4 + 4 \div 4 = 2$; $(4 + 4 + 4) \div 4 = 3$

Part 2
- For this activity the class needs coloured card or gummed paper and scissors.
- Ask the class to cut out two trapezia (plural of trapezium; can also be trapeziums) that are the same size, and arrange them to form a parallelogram by rotating one by half a turn:

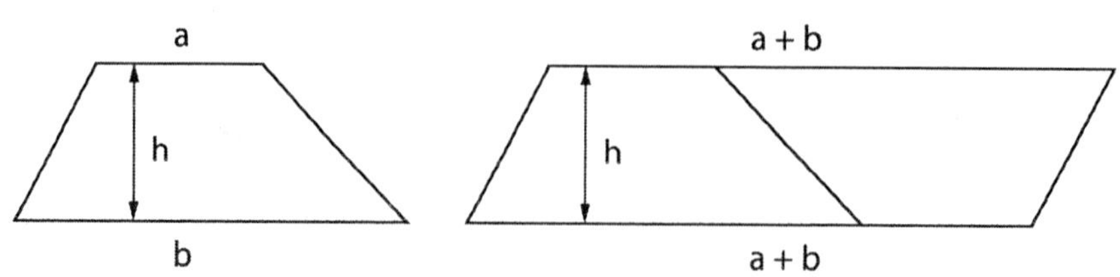

- This shows that the area of the trapezium is half the area of a parallelogram. So the area of a trapezium is $\frac{1}{2}$ × sum of the lengths of the parallel sides × the height.
- The area of a trapezium is given by the formula: $A = \frac{1}{2} \times (a + b) \times h = \frac{1}{2}(a + b)h$

- Show the class the example below. On the diagram, $a = 7$ cm, $b = 3$ cm and $h = 4$ cm. The area is: $A = \frac{1}{2} \times (3 + 7) \times 4 = \frac{10 \times 4}{2} = 20$ cm^2

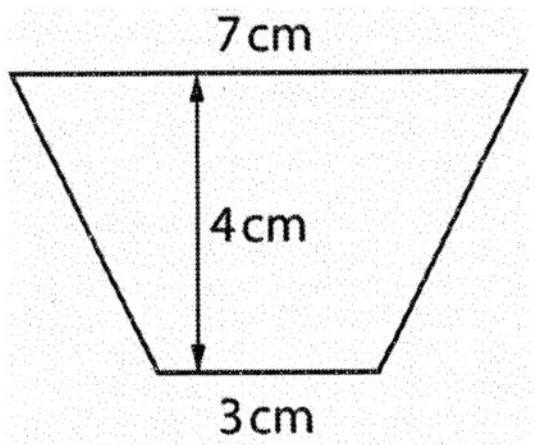

- **Pupils can now do Exercise 6C from Pupil Book 2.2.**

Part 3

- This is a revision exercise to make sure that pupils know the formulae for the area of different quadrilaterals.
- Draw the following quadrilaterals on the board or OHP: a square, a rectangle, a triangle, a parallelogram and a trapezium.
- Ask pupils to write down in their books the formula to find the area of each shape.
- To check answers, ask individuals to write the formula for each quadrilateral on the board.

Answers

Exercise 6C

1 **a** 72 cm^2 **b** 25 cm^2 **c** 3m^2 **d** 22m^2 **e** 325mm^2

2 **a** 15 cm^2 **b** 66 cm^2 **c** 30m^2 **d** 4cm **e** 10cm **f** 10m

3 30m^2

4 56cm^2

5 44 cm^2

6 $a = 6$ cm, $b = 14$ cm

7 75 cm^2

8 72cm^2

9 8 cm

10 Values of a, b and h with $(a + b) \times h = 12$ and $b > a$
(for example, $a = 1$, $b = 5$, $h = 2$; $a = 2$, $b = 4$, $h = 2$; $a = 1$, $b = 2$, $h = 4$)

Problem solving: Pick's formula

A

Shape	Number of dots on perimeter of shape	Number of dots inside shape	Area of shape (cm^2)
a	8	1	4
b	12	3	8
c	8	3	6
d	4	2	3
e	9	4	7.5
f	10	4	8
g	11	3	7.5
h	14	4	10

B $A = \frac{1}{2}P + I - 1$

C Check pupils' shapes

Lesson 6.4 Surface areas of cubes and cuboids

Learning objective

- To find the surface areas of cubes and cuboids

Resources and homework

- Pupil Book 2.2, pages 112–115
- Homework Book 2, section 6.4
- Online homework 6.4, questions 1–10

Links to other subjects

- **Design and technology** – to design packaging material
- **Art** – to design 3D sculptures involving cubes and cuboids

Key words

- cube
- surface area
- cuboid

Problem solving and reasoning help

- **PS** question 8 of Exercise 6D in the Pupil Book helps pupils to extend their understanding of cubes and cuboids to compound 3D shapes. If pupils struggle, provide concrete examples to help them visualise what they need to do. The investigation at the end of the exercise encourages pupils to think about the classic problem of maximising the volume for a given surface area. In this example, pupils need to use their understanding of sequences to generalise the relationship between the total surface area and the component parts. Again, provide pupils who struggle with concrete examples to explore. Encourage **more able** pupils to try to explain why any rules they find make sense, based on the structure of the problem.

Common misconceptions and remediation

- Pupils often confuse the concept of surface area and volume. Using concrete examples will help pupils to understand the difference.

Probing questions

- How do you find the surface area of a cuboid?
- Find three cuboids with a surface area of 48 cm^2.

Part 1

- Write the number 12 on the board.
- Ask individuals to come up and write on the board three numbers that have a product of 12, allowing repeats. Remind pupils that 'product' means 'multiply'.
- Examples are: 1 × 1 × 12, 1 × 2 × 6, 1 × 3 × 4, 2 × 2 × 3.
- Repeat the activity using different numbers.

Part 2

- Show the class a cuboid made from multi-link cubes with:
  length = 4 cm, width = 3 cm and height = 2 cm.
- Remind the class how to find the total surface area of the cuboid by finding the area of its six surfaces and adding them together. Draw the cuboid on the board.
- The formula to find the total surface area of any cuboid is:
  $A = 2lw + 2lh + 2wh$
- So the surface area of the multi-link cuboid described is:
  $A = (2 \times 4 \times 3) + (2 \times 4 \times 2) + (2 \times 3 \times 2) = 24 + 16 + 12 = 52\ cm^2$
- Explain to the class that *volume* is the amount of space

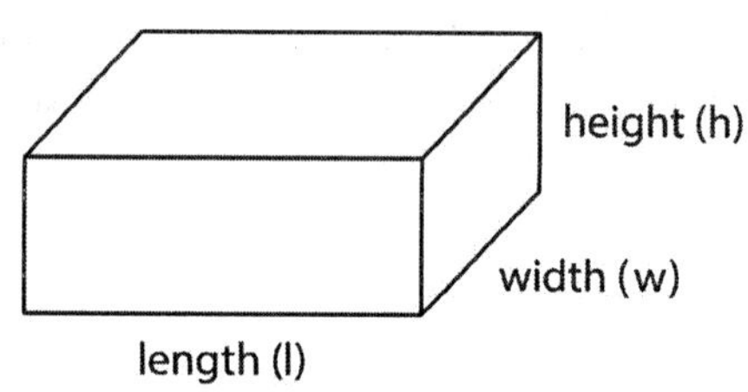

inside a 3D shape.

- Show the class the multi-link cube and explain that the volume is made from 24 cubes.
- A quick way to find the volume of a cuboid is to multiply its length by its width by its height.
- Using the same diagram: The volume of a cuboid = length × width × height
- $V = l \times w \times h = lwh$
- The metric units of volume in common use are: the cubic millimetre (mm^3), the cubic centimetre (cm^3), the cubic metre (m^3).
- So the volume of the multi-link cuboid above is: $V = 4 \times 3 \times 2 = 24$ cm^3.
- Explain to pupils that the *capacity* of a 3D shape is the volume of liquid or gas it can hold. The metric unit of capacity is the litre (l). Pupils should write down the following:
- 100 centilitres (cl) = 1 litre, 1000 millilitres (ml) = 1 litre
- Show the class various objects for which capacity is used, such as a 5 ml spoon or a 70 cl wine bottle.
- The following metric conversions for capacity should also be noted:
- 1 litre = 1000 cm^3, 1 ml = 1 cm^3, 1000 litres = 1 m^3.
- **Pupils can now do Exercise 6D from Pupil Book 2.2.**

Part 3

- Draw a cuboid on the board or OHP as shown.
- Ask the class to explain how to find the total surface area of the cuboid.
- Ask the class to explain how to find the volume of the cuboid.

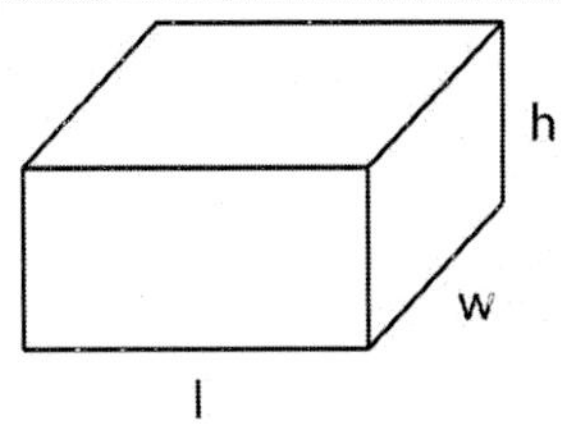

Answers

Exercise 6D

1 **a** 208cm^2 **b** 164cm^2 **c** 290 cm^2 **d** 40cm^2

2 214cm^2

3 **a** 96cm^2 **b** 216cm^2 **c** 384cm^2 **d** 486cm^2

4 **a** 6cm^2 **b** 150cm^2 **c** 600cm^2 **d** 864cm^2

5 1238cm^2

6 62m^2

7 16 400cm^2

8 44m^2

Investigation: An open box problem

A

Size of square cut off	Area of the four squares	Surface area of box
1cm by 1cm	4cm^2	248cm^2
2cm by 2cm	16cm^2	236cm^2
3cm by 3cm	36cm^2	216cm^2
4cm by 4cm	64cm^2	188cm^2
5cm by 5cm	100cm^2	152cm^2
6cm by 6cm	144cm^2	108cm^2

B The width of the card would be 0

C **a** 12, 20, 28, 36, 44 **b** add 8 **c** $8n + 4$

Review questions (Pupil Book pages 116–117)

- The review questions will help to determine pupils' abilities with regard to the material within Chapter 6.
- These questions also draw on the mathematics covered in earlier chapters of the book to encourage pupils to make links between different topics.
- The answers are on the next page.

Investigation – A cube investigation (Pupil Book pages 118–119)

- Pupils apply their understanding of area to a more complex problem.
- Pupils need to work methodically and be able to explain their solutions. This is a good transferable skills objective to share with pupils when doing this investigation. Ask pupils to share not only their solutions but also *how* they approached working on the problem.
- Pupils should work in small groups to encourage discussions about the problem.
- You could provide **less able** pupils with linking cubes to help them visualise the problem.
- You could use strategies such as reading images to structure pupils' responses to the final questions. For example:
    - What is the least surface area for the different shapes you have made? Explain why.
    - What do you think the surface areas of a 3D shape made from five cubes could be?
    - Work out the least surface area of a 3D shape made from six cubes.
- Pupils should write these questions in the centre of a large sheet of paper. Around the outside, give fixed sub-questions to which they should respond, for example, 'What do you know? What does this mean? Can you see any patterns? Can you explain these patterns?' This will scaffold pupils' thinking about the problem.
- You could ask **more able** pupils to provide more detailed justifications of any rules, by revisiting the structure of the problem.

Answers to Review questions

1 **a** Pack of 24 **b** $1300cm^2$

2 **a** For example:

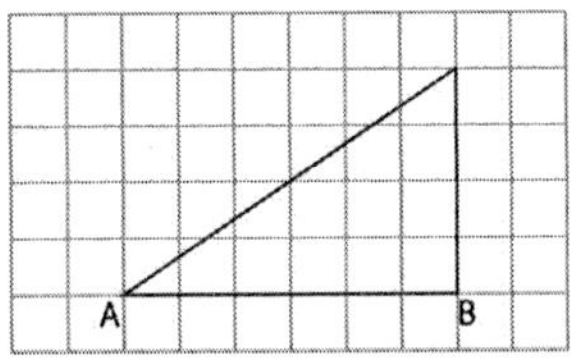

b

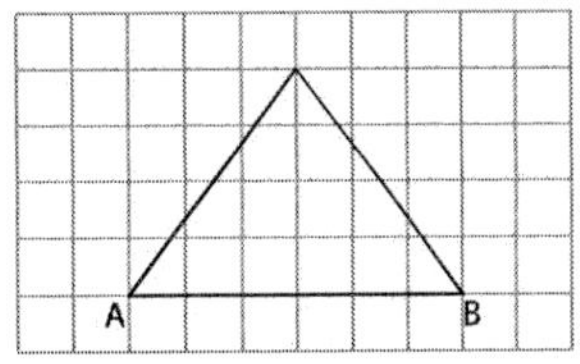

3 **a** $31.5cm^2$ **b** $25cm^2$ **c** $12m^2$

4 $24cm^2$

5 **a** $13.5m^2$ **b** $70cm^2$ **c** $60cm^2$

6 **a** $l = 5cm$; $h = 4cm$; $w = 3cm$ **b** $94cm^2$

Answers to Investigation – A cube investigation

1

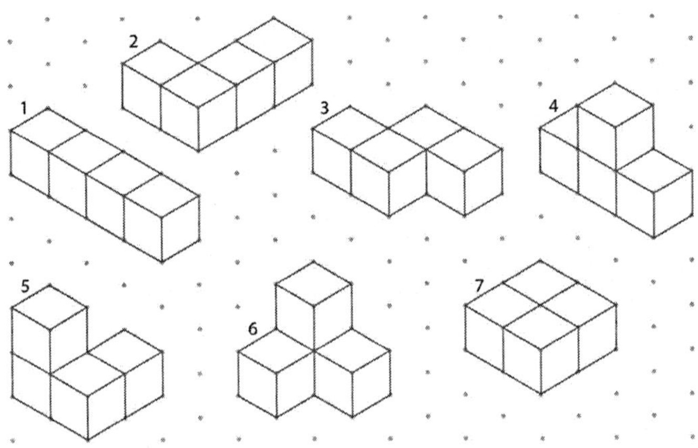

2

3D shape	1	2	3	4	5	6	7
Surface area	$18cm^2$	$18cm^2$	$18cm^2$	$18cm^2$	$18cm^2$	$18cm^2$	$16cm^2$

3 Shape 7 has the least surface area and the rest have the same surface area. The shape with the least surface area has four pairs of faces touching, so leaving 16 faces exposed. The other six have three pairs of faces touching, so leaving 18 faces exposed.

4 A shape made from five cubes must have four or five pairs of faces touching, so the surface areas are either $20cm^2$ or $22cm^2$.

For all the shapes in this investigation, the surface area is an even number of square centimetres.

Two cubes have 12 faces in total, so if one pair of faces is touching, then 10 faces are exposed.

Three cubes have 18 faces in total, so if two pairs of faces are touching, then 14 faces are exposed.

A shape made from four cubes must have three or four pairs of faces touching, so either 16 or 18 faces are exposed.

7 Graphs

Learning objectives

- How to draw graphs of linear equations
- How to work out the gradient of a linear graph
- How to work out an equation of the form $y = mx + c$ from its graph
- How to draw graphs of simple quadratic equations
- How to draw graphs to illustrate real-life situations.

Prior knowledge

- How to plot coordinates in all four quadrants
- How to calculate with negative numbers

Context

- This chapter builds on previous work on mapping diagrams and graphs covered in Year 7, where pupils identified functions from inputs and outputs (including the inverse function) and related these to coordinate pairs, which are used to draw graphs. The important concept of the gradient of a straight line is introduced in this chapter and pupils will explore the form $y = mx + c$ for a straight line. Pupils are introduced to the concept of the gradient of a straight line and to graphs of quadratic functions.

Discussion points

- Without drawing the graphs, compare properties of pairs of graphs such as:

  $y = 2x$
  $y = 2x + 4$
  $y = x + 5$
  $y = x - 2$
  $y = 4x - 3$
  $y = -4x + 4$

Associated Collins ICT resources

- Chapter 7 interactive activities on Collins Connect online platform
- *Skiing* Wonder of Maths on Collins Connect online platform

Curriculum references

Develop fluency

- Move freely between different numerical, algebraic, graphical and diagrammatic representations [for example, equivalent fractions, fractions and decimals, and equations and graphs]
- Develop algebraic and graphical fluency, including understanding linear and simple quadratic functions

Reason mathematically

- Extend their understanding of the number system; make connections between number relationships, and their algebraic and graphical representations

Algebra

- Substitute numerical values into formulae and expressions, including scientific formulae
- Model situations or procedures by translating them into algebraic expressions or formulae and by using graphs4
- Work with coordinates in all four quadrants
- Recognise, sketch and produce graphs of linear and quadratic functions of one variable with appropriate scaling, using equations in x and y and the Cartesian plane
- Reduce a given linear equation in two variables to the standard form $y = mx + c$; calculate and interpret gradients and intercepts of graphs of such linear equations numerically, graphically and algebraically
- Find approximate solutions to contextual problems from given graphs of a variety of functions, including piece-wise linear, exponential and reciprocal graphs

Fast-track for classes following a 2-year scheme of work

- It is important to take time over the examples in this chapter.
- Sometimes, however, it is more worthwhile to work through one or two examples in depth as a class, followed by picking out just one or two key examples for pupils.

Lesson 7.1 Graphs from linear equations

Learning objective

- To recognise and draw the graph of a linear equation

Resources and homework

- Pupil Book 2.2, pages 121–123
- Homework Book 1, section 7.1
- Online homework 7.1, questions 1–10

Links to other subjects

- **Science** – to use and manipulate scientific formulae

Key words

- linear equation
- variable

Problem solving and reasoning help

- The **PS** questions (6e and 7g) and the challenge at the end of Exercise 7A of the Pupil Book encourage pupils to make links between the gradient and intercepts of graphs based on their equations. These questions will give pupils a better understanding when they meet this topic in Lesson 7.2.

Common misconceptions and remediation

- Pupils often struggle to recognise that letters represent variables and that the answer can vary depending on the situation. Provide lots of opportunity for pupils to see this in action in contexts with which they are familiar, for example, 'Think of a number' word problems. Pupils need to understand that letter symbols used in algebra stand for unknown numbers or variables and not labels. For example, $5b$ cannot mean 5 bananas.

Probing questions

- How do you find a set of coordinates for a straight line graph such as: $y = 2x + 4$?
- How do you decide on the range of numbers to put on the x and y axes?
- How do you decide on the scale you are going to use?

Part 1

- On the board, write the following puzzle:

```
 CATS
 HATE
 ----
 DOGS
```

NB one possible solution:

```
 4317
 5310
 ----
 9627
```

- Tell pupils that each letter stands for a different number. Ask pupils what they know about certain letters. For example, E = 0, both C and H are less than D, G must be even, and there are a few more sensible observations.
- Ask for a suggestion for C, H and D, then follow this suggestion through and see if it is a possible solution. There are quite a few solutions.
- Once pupils have suggested a correct solution (which may be after a few wrong attempts), ask them to work in pairs or groups of three or four to suggest another possible solution.

Part 2

- Discuss the way we write functions, for example: $x \rightarrow 4x + 3$. Give some other examples.
- Explain that we can write functions in different ways, for example: $y = 4x + 3$, which means the same thing as $x \rightarrow 4x + 3$.
- This is a simpler way of looking at the function if we want to draw a graph from it.
- We can draw a graph of every function; the coordinates come from the combination of inputs and outputs. Tell pupils that they are going to create some of these graphs.
- On the board, write $x \rightarrow 3x + 1$ and show how we can work with $y = 3x + 1$, as the y simply tells us the output for different input values of x.
- Write this table on the board:

x	−2	−1	0	1	2	3
y						

- Using this, we can create a table of values that combines the inputs and outputs.
- Show how we use the function $x \rightarrow 3x + 1$ to substitute the different values of x as the inputs to find the outputs y. We can use any values for x that we want, but it is a good idea to use those around 0.
- Suggest that it is best to start with the values 0, 1, 2 and 3 before moving to negative values.
- Go through the substitution with the class to give the values in the table as shown:

x	–2	–1	0	1	2	3
y	–5	–2	1	4	7	10

- To draw the graph, we need a pair of axes that use all the values in the table, in other words: x from –2 to 3 and y from –5 to 10.
- Draw the axes on the board, showing the linear scale, and add all the numbers. Remind pupils how to plot the coordinates from the table and join them with a straight line.
- Point out that the straight line represents ALL the points that satisfy $y = 3x + 1$. Demonstrate this by choosing any non-integer point such as $x = 1.5$, go through the calculations $(1.5 \times 3) + 1 \rightarrow 5.5$ and show that (1.5, 5.5) is on the line.
- Show that this also works for a few other points such as: $x = 2.2$ (7.6) and $x = -0.5$ (–0.5).
- **Pupils can now do Exercise 7A from Pupil Book 2.2.**

Part 3

- On the board, write: $y = 5x + 17$ and $y = 5x + 11$.
- Ask what pupils can tell you about the graphs of these two functions.
- They should be able to tell you that the lines have the same slope (are parallel), and that the first line intercepts the y-axis at a higher number than the second line.

Answers

Exercise 7A

1 a 1, 2, 3, 4, 5, 6 **b** and **c** Graph with straight line through (–2, 1) and (3, 6)

2 a –4, –3, –2, –1, 0, 1 **b** and **c** Graph with straight line through (–2, –4) and (3, 1)

3 a –6, –3, 0, 3, 6, 9 **b** and **c** Graph with straight line through (–2, –6) and (3, 9)

d

–2	–1	0	1	2	3
–4	–2	0	2	4	6
–8	–4	0	4	8	12

f All straight and pass through (0,0)
g i Graph with straight line through (–2, –5) and (3, 7.5) **ii** graph with straight line through (–2, –1) and (3, 1.5)

4 a

–8	–4	0	4	8	12
–7	–3	1	5	9	13

b and **c** Graph with straight line through (–2, –7) and (3, 13)

5 a

–8	–4	0	4	8	12
–9	–5	–1	3	7	11

b and **c** Graph with straight line through (–2, –9) and (3, 11)

6 a

3	4	5	6	7	8
1	2	3	4	5	6
–1	0	1	2	3	4
–3	–2	–1	0	1	2
–5	–4	–3	–2	–1	0

d All straight and parallel to each other

e i Graph with straight line through (–2, 0.5) and (3, 5.5)
ii Graph with straight line through (–2, –3.5) and (3, 1.5)

7 a

–4	–2	0	2	4	6
0	2	4	6	8	10

c Graph with straight line through (–2, 0) and (3, 10)

d

–4	–2	0	2	4	6
–2	0	2	4	6	8
–6	–4	–2	0	2	4
–8	–6	–4	–2	0	2

f All straight and parallel to each other
g i Graph with straight line through (–2, –1.5) and (3, 8.5)
ii Graph with straight line through (–2, –6.5) and (3, 3.5)

Challenge: Sloping graphs
Graph opposite

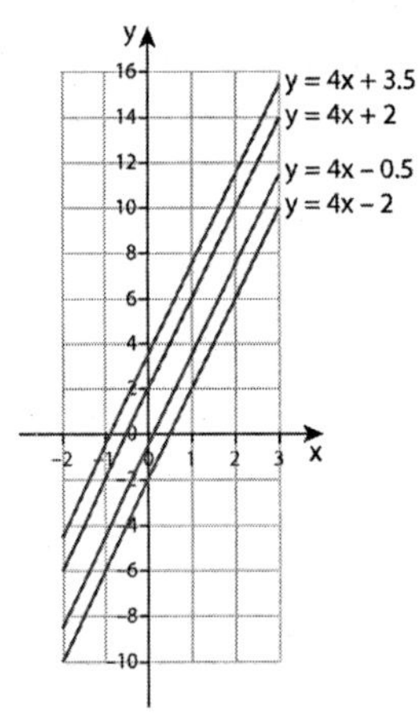

Lesson 7.2 Gradient (steepness) of a straight line

Learning objective

- To work out the gradient in a graph from a linear equation
- To work out an equation of the form $y = mx + c$ from the graph

Resources and homework

- Pupil Book 2.2, pages 124–126
- Homework Book 2, section 7.2
- Online homework 7.2, questions 1–10

Links to other subjects

- **Science** – to use and manipulate scientific formulae

Key words

- constant
- gradient
- *y*-intercept

Problem solving and reasoning help

- The **PS** question 7 in Exercise 7B of the Pupil Book requires pupils to write equations of the form $y = mx + c$ from given coordinate points on the line. Encourage **less able** pupils to draw the line and find the intercept and gradient, whereas you could ask **more able** pupils to calculate the gradient from the given points. The challenge question asks pupils to find as many equations as possible that pass through a single coordinate point on the line. This task encourages pupils to be more fluent and confident in their use of algebra and to realise that they should not always be looking for 'one correct' answer.

Common misconceptions and remediation

- Pupils often ignore the scales on the axes and just count the squares when calculating the gradient or divide the change in x over the change in y by mistake. Pupils will also occasionally give the *x*-intercept instead of the *y*-intercept. Make certain these points are addressed and pupils have a clear understanding of how to do these questions correctly.

Probing questions

- If you increase or decrease the value of m, what effect does this have on the graph? What about changes to c?
- What have you noticed about the graphs of functions of the form $y = mx + c$? What elements are similar? And what elements can be different?

Part 1

- Ask the class if anyone can tell you a shortcut way to multiply by 29. Someone should suggest multiplying by 30, and then subtracting one of the numbers by which you are multiplying. Pupils may come up with other valid suggestions.
- Ask if anyone can multiply 43 by 29:
- 43 × 30 = (129 × 10) and then 1290 – 43 = 1247 (Ask a pupil to check this on a calculator.)
- Ask if anyone can multiply 24 by 29:
- 24 × 30 = (72 × 10) and then 720 – 24 = 696 (Ask a pupil to check this on a calculator.)
- Ask if anyone is up for a three-digit challenge, to multiply a three-digit number by 29. A volunteer should multiply 147 by 29: 147 × 30 = (441 × 10); then 4410 – 147 = 4263.

Part 2

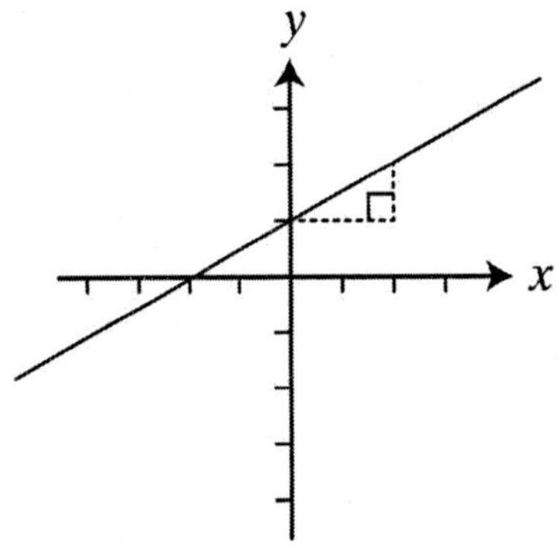

- Draw a straight line on the board, slanting up towards the top right-hand corner.
- Ask: 'How steep is this line?' Answers such as 'very steep', or 'quite steep' should lead the class to recognise that steepness needs to be measured in some way.
- Now add to the diagram a pair of axes around the line, so that the line intercepts the y-axis at a positive integer. Label the axes with suitable numbers.
- Add a right-angled triangle with a difference of 1 along the *x*-axis, as shown on the diagram.
- The gradient is 'the increase in the *y*-coordinate for an increase of 1 in the *x*-coordinate'.
- Find the increase in the *y*-coordinate on the triangle; this is the gradient, which is the measure of steepness.
- Every straight line has a gradient that can be found as shown here. Show pupils how to find the gradient of a steeper line, pointing out that the gradient is a larger number.
- Explain that every linear function gives a straight-line graph of the form $y = mx + c$,
- Where m is the gradient and c is the value at which the *y*-axis is intercepted.
- You have two straight lines around a pair of axes on the board. Use these to illustrate how to find the equation of these lines by using the gradient for m and the *y*-axis intercept as c.
- Draw another pair of labelled axes with a straight line (positive gradient).
- Ask pupils what the gradient is, and go through the procedure (gives m). Ask pupils where the line intercepts the *y*-axis (gives c). Write down the equation of this straight line.
- **Pupils can now do Exercise 7B from Pupil Book 2.2.**

Part 3

- Ask if anyone can tell you what a gradient is. It is a measure of slope, or how much up you have gone for one unit along.
- On the board, write the equation: $y = 6x + 5$. Ask what the gradient of the graph is. (It is 6.)
- Ask whereabouts the graph intercepts the *y*-axis. (It intercepts at 5.)
- Repeat for a few equations more until everyone can give the correct answer. Ask pupils if the graph drawn from something like $y = 8x + 17$ will always be a straight line. (Yes)

Answers

Exercise 7B

1 **a** 3 **b** 2 **c** 4 **d** 1

2 **a** $y = 3x + 5$ **b** $y = 2x + 7$ **c** $y = x + 4$ **d** $y = 7x + 15$

3 **a** (0, 1) **b** 2 **c** $y = 2x + 1$

4 **a i** 2 **ii** (0, 3) **iii** $y = 2x + 3$ **b i** 3 **ii** (0, 1) **iii** $y = 3x + 1$
c i 4 **ii** (0, 2) **iii** $y = 4x + 2$ **d i** 2/3 **ii** (0, 3) **iii** $y = x + 3$

5 **a** (0, 4) **b** 2 **c** $y = 2x + 4$

6 **a** 3 **b** (–1, 0) **c** $y = 3x – 1$

7 **a** $y = 3x + 2$ **b** $y = 4x – 1$ **c** $y = 2x + 1$

Challenge: Lines through points

a For example $y = x + 4$, $y = 2x + 4$, $y = 3x + 4$

b For example $y = x$, $y = 2x – 1$, $y = 3x – 2$

c For example $y = x + 3$, $y = 2x + 1$, $y = 3x – 1$

Lesson 7.3 Graphs from simple quadratic equations

Learning objective

- To recognise and draw the graph from a simple quadratic equation

Resources and homework

- Pupil Book 2.2, pages 126–129
- Homework Book 1, section 7.3
- Online homework 7.3, questions 1–10

Links to other subjects

- **Science** – to graph data from experiments such as throwing a ball
- **Economics** – to work with profit curves

Key words

- quadratic

Problem solving and reasoning help

- **PS** question 7 in Exercise 7C of the Pupil Book requires pupils to transfer their knowledge of drawing quadratic graphs to a real-life situation. Offer guidance to **less able** pupils, which may include setting up and completing the table.

Common misconceptions and remediation

- Pupils will often misread the scales when reading or drawing quadratic graphs. Another common problem relates to errors in calculations which stem from an incorrect understanding of BIDMAS. Pupils will often draw the bottom of a quadratic graph flat if its minimum point lies between two plotted points.

Probing questions

- How can you tell if a function is a quadratic function from its equation?
- How do you find coordinates for a quadratic function?
- Are there any coordinates on the graph of $y = 2x^2 + 3$, which lie below the x-axis?
- Do all quadratic graphs have lines of symmetry? Explain why.

Part 1

- Ask pupils to square the numbers from 0 to 10. Ask **more able** pupils to square negative numbers from −1 to −10.

Part 2

- Explain that quadratic graphs are always curved and that they always have an x^2 in their equation. Then explain that drawing a quadratic graph is similar to drawing a linear graph: As you complete a table of coordinate values, plot these points on a set of suitable axes and join the points with a smooth curve.
  Go through the example below. Copy and complete this table for the equation $y = x^2 + 3$.

x	−3	−2	−1	0	1	2	3
x^2	9	4	1	0	1	4	9
$y = x^2 + 3$	12	7	4	3	4	7	12

- Emphasise that the x values are always given in questions of this type.
- Explain that the maximum and minimum values of the axes are generated by the maximum and minimum values of x and y (minimum of 0 for y if all y values are positive as in the example). In this case draw the x-axis from –3 to +3 and the y-axis from 0 to 12.

- Ask pupils why we used 0 as the starting point for the y-axis and not 3.
- Draw the graph from the table for pupils. Show pupils the line of symmetry on the graph.
- **Pupils can now do Exercise 7C from Pupil Book 2.2.**

Part 3

- Display graphs of:
  - $y = x^2 + 1$
  - $y = x^2 - 1$
  - $y = 2x^2$
  - $y = -x^2$
  - $y = -x^2 + 1$
  - $y = -x^2 - 1$
- Ask pupils to comment on the x- and y-intercepts the maximum and minimum values and the symmetry of the graphs.

Answers

Exercise 7C

1 **a** 10, 5, 2, 1, 2, 5, 10 **b** and **c** Check pupils' graphs

2 **a**

9	4	1	0	1	4	9
12	7	4	3	4	7	12

b and **c** Check pupils' graphs

3 **a**

11	6	3	2	3	6	11
13	8	5	4	5	8	13
14	9	6	5	6	9	14

b, **c** and **d** Check pupils' graphs

d each line is curved and parallel to each other

4 **a**

18	8	2	0	2	8	18

b and **c** Check pupils' graphs

5 **a**

9	4	1	0	1	4	9
27	12	3	0	3	12	27

b and **c** Check pupils' graphs

6 **a**

36	16	4	0	4	16	36
45	20	5	0	5	20	45
54	24	6	0	6	24	54

b, **c** and **d** Check pupils' graphs

d each one is a curve with the graph getting more squashed as the number next to the x gets larger

7 **a** Check pupils' graphs **b** 18 km **c** 44 km

Lesson 7.4 Real-life graphs

Learning objective

- To draw graphs from real-life situations to illustrate the relationship between two variables

Links to other subjects

- **PE** – working with time, distance and average speed.
- **Science** – distance–time graphs

Resources and homework

- Pupil Book 2.2, pages 129–131
- Homework Book 1, section 7.4
- Online homework 7.4, questions 1–10

Key words

- average speed
- distance–time graph

Problem solving and reasoning and financial skills help

- Each **PS** question in Exercise 7D of the Pupil Book involves reading values from distance–time graphs. Emphasise that accuracy when drawing these graphs is very important if they are to produce correct answers. The **MR** questions in the exercise require pupils to explain how to calculate average speed and how to read from a graph. **Less able** pupils may need support in order to structure coherent answers.

Common misconceptions and remediation

- Pupils can get confused between the different ways that times are written. For example, they may assume that 1.4 hours is the same as 1 hour and 40 minutes, or that 1 hour and 50 minutes is 1.5 hours. Make these points to pupils during the lesson

Probing questions

- Show pupils a travel graph with two journey lines that cross at some point.
- What does the gradient of the line tell us?
- What do the flat parts tell us?
- What does it mean when the graphs cross?

Part 1

- Draw a circle on the board, and write the numbers 36, 25, 7, 3 and 4 inside the circle.
- Also write the number 518 in a box as a target.
- Ask pupils to suggest a combination of some or all of the numbers in the circle, together with the four operations and brackets that will give a total as close to the target number as possible. This works well with pupils working in pairs, or small groups of three or four pupils.
- If you do have specific evenly-balanced teams, you can play a few rounds and keep a score, for example, spot-on scores 10, closest scores 5, next closest 4 and so on down to 0.
- In each round, choose different targets and change the numbers in the circle.
- The beauty of this game is that you do not need to know the numbers beforehand, as the target number does not have to be achieved exactly.

Part 2

- Ask pupils where they see graphs the most (for example, television, papers, adverts).
- Some of these may be what we call travel graphs or time–and–distance graphs.
- Talk pupils through Example 5 in the Pupil Book (but do not use the textbook). Tell the story about (you or a friend) picking up a dog from the vet. Write key information on the board:
  - Travel $1\frac{1}{2}$ hours at 60 km/h, so that is 90 km;

- Took 30 minutes to settle the dog, so no distance travelled in that time;
- Came back at 40 km/h (gently, for the benefit of the dog), which took $2\frac{1}{4}$ hours.

- Discuss the axes needed – the vertical axis is 'the distance from home'; the horizontal axis is 'time'. Draw the pair of axes on the board and label them, with the vertical axis from 0 to 100 and the horizontal axis from 0 to 5.
  Key coordinates need to be identified to draw the graph: Leave from home (0, 0); arrive at the vet ($1\frac{1}{2}$, 90); leave the vet (2, 90); arrive at home ($4\frac{1}{4}$, 0).
- Draw the graph of the journey, joining point to point with straight lines.
- You will need to explain that the straight lines are to illustrate the average speed travelled (the car did not actually stay at one speed all the time).
- **Pupils can now do Exercise 7D from Pupil Book 2.2**

Part 3

- Ask what average speed pupils walk at to school (for the parts of their journey that they do walk). This is usually between three and four miles an hour, unless they dawdle.
- Ask pupils if they actually walk at the same speed all the time. Do they walk quicker uphill or downhill? The notion of average speed is exactly what it says, a rough middle-of-the-line speed, which is why it is used on travel graphs.
- Ask a pupil to draw what might be called 'A real travel graph of walking at 4 mph'. This would be a wiggly line that meanders around the straight line.
- Ask for other attempts at a real average 4 mph.
- Discuss with the class why we use one straight line to represent the average speed.

Answers

Exercise 7D

1 **a** and **b** Check pupils' graphs **c** 2 pm
2 **a** and **b** Check pupils' graphs **c** $S = D \div T = 20 \div \frac{1}{3} = 60$ km/h
3 **a** Check pupils' graphs **b** 0, 40, 100, 160, 200 **c** 70 cm
d Check pupils' explanations
4 **a** Check pupils' graphs **b** 5:10 pm
5 **a** 7400, 5800, 4200, 2600, 1000, X
b Check pupils' graphs
c About 280 minutes (4 hours 40 minutes)

Problem solving: Meeting in the middle

A Check pupils' graphs
B **a** About 12:17 pm **b** About 11:48 am

Review questions

(Pupil Book pages 132–133)

- The review questions will help to determine pupils' abilities with regard to the material within Chapter 7.
- These questions also draw on the maths covered in earlier chapters of the book to encourage pupils to make links between different topics.
- The answers are on the next page.

Challenge – The M25

(Pupil Book pages 134–135)

- This challenge activity encourages pupils to think about the M25, one of Europe's busiest motorways. Read and then discuss the text at the top of page 134 of the Pupil Book. Ask pupils some questions relating to the text. Topics could include:
  - reading the map (scale, what roads intersect, why some roads are blue)
  - conversion factors (imperial to metric)
  - average traffic flow (per hour, per year).
- Encourage pupils to suggest possible questions. They could do these in small groups or individually, then present these to the class for others to answer. This is particularly useful if pupils have access to the internet.
- Ask the class to work through questions 1–8 in the challenge activity. You may need to discuss some of the mathematics and strategies needed to work out some of the questions.
- You could extend this challenge activity by giving pupils the following question. The table shows the lengths of the three longest orbital motorways in the world and the length of the M60, which is the only other orbital motorway in Britain.

City	Country	Road	Length (miles)
Berlin	Germany	B-10	122
London	England	M25	117
Cincinnati	USA	I-275	84
Manchester	England	M60	35

The formula for the radius of a circle given the circumference is $r = \frac{C}{2\pi}$.

- Assuming that the M25 is a circle with a circumference of 117 miles, what would the radius be? (18.6 miles)
- Making the same assumption calculate the radii of the other orbital roads. (B-10 19.4 miles, I-275 13.4 miles, M60 5.6 miles)

Answers to Review questions

1 **a** Ground (0), and 20 **b** 62 seconds **c** Straight line from (70, 20) to (105, 0)
2 **a** 19 **b** –11 **c** $y = x^2$
3 **a** 17 cm **b** 450 cm²
4 **a** $y = x + 3$ **b** $y = x – 3$ **c** $y = –x + 3$ **d** $y = –x – 3$
5 **a** Check pupils' graphs **b** 56 cm **c** 1.7 seconds

Answers to Challenge – the M25

1 14 years
2 **a** 31 **b** 5 **c** Surrey
3 **a** 13 miles **b** 20 miles
4

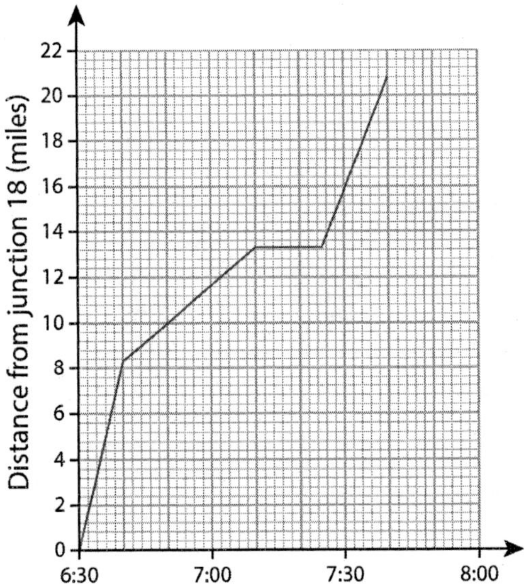

5 **a** 732 miles
b

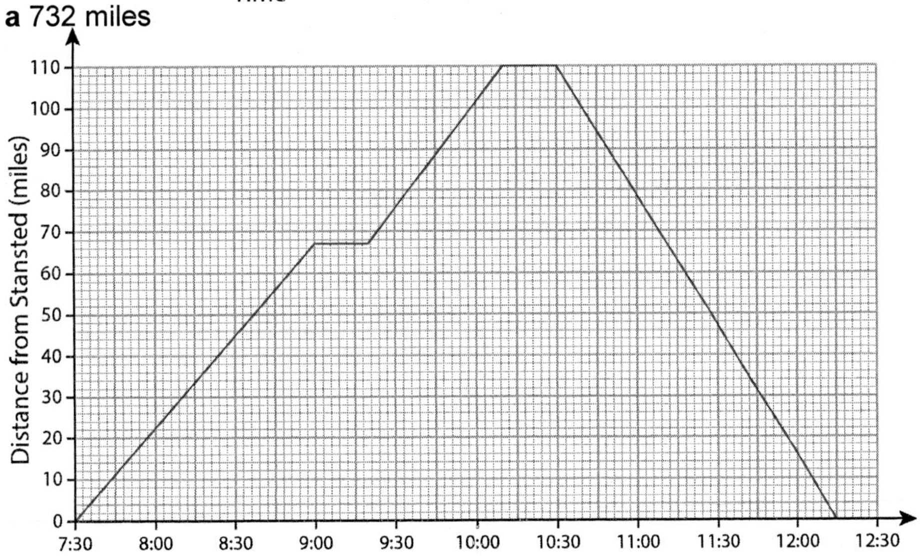

6 1 hour 40 minutes
7 187 km
8 19%

8 Simplifying numbers

Learning objectives

- How to multiply and divide by powers of 10
- How to round numbers to a specific number of significant figures
- How to write large numbers in standard form
- How to multiply numbers in standard form

Prior knowledge

- How to multiply and divide by 10, 100 and 1000
- How to round to the nearest 10, 100 and 1000

Context

- This chapter builds on previous work with decimals, introducing powers of 10 as a lead-in to working with standard index form. Estimation is used as a means to test whether or not answers are realistic or sensible. Some of the work is specifically designed to reinforce pupils' mental arithmetic skills; other work is about using calculators efficiently. You could introduce standard form as a powerful tool that is widely used in science, especially by scientists dealing with massive spaces. For example, our nearest star, Proxima Centauri, is 40 653 234 200 000 kilometres from Earth. An atom is 0.000 000 0001 metres wide. You could search on the internet for 'standard form' or 'very large and very small numbers', or use some of the problems on the STEM Nrich website to extend pupils' experience of very large and very small numbers: **http://nrich.maths.org/stemnrich**

Discussion points

- Explain whether the following are true or false:
    - 4.399 rounds to 4.310 to two decimal places (false)
    - 6.5999 rounds to 6.6 to one decimal place (true).
- How do you go about rounding a number to one decimal place?
- Why might it not be possible to identify the first three places in a long jump competition if measurements were taken in metres to one decimal place?
- Show me a length that rounds 5.6 m to one decimal place. Are there other lengths?

Associated Collins ICT resources

- Chapter 8 interactive activities on Collins Connect online platform
- *Understanding standard form* video on Collins Connect online platform
- *Pyramids* Wonder of Maths on Collins Connect online platform

Curriculum references

Develop fluency

- Consolidate their numerical and mathematical capability from key stage 2 and extend their understanding of the number system and place value to include decimals, fractions, powers and roots

Solve problems

- Develop their use of formal mathematical knowledge to interpret and solve problems, including in financial mathematics
- Begin to model situations mathematically and express the results using a range of formal mathematical representations

Number

- Interpret and compare numbers in standard form *A* x 10*n* 1 ≤ *A* < 10, where n is a positive or negative integer or 0
- Round numbers and measures to an appropriate degree of accuracy [for example, to a number of decimal places or significant figures]

Fast-track for classes following a 2-year scheme of work

- There are new ideas in all these chapters but they do build on pupils' existing knowledge of rounding and the number system. Check understanding by doing a couple of examples as a class, and then ask pupils to focus on the **PS** and **MR** questions in the exercises, plus the investigations, activity and challenge at the end of the exercises in this chapter.

Lesson 8.1 Powers of 10

Learning objective

- To multiply and divide by powers of 10

Resources and homework

- Pupil Book 2.2, pages 137–138
- Homework Book 2, section 8.1
- Online homework 8.1, questions 1–10

Links to other subjects

- **Science** – to describe scientific phenomena

Key word

- power

Problem solving and reasoning help

- The **MR** question 8 and **PS** question 9 in Exercise 8A of the Pupil Book requires pupils to apply their understanding to engaging but less familiar situations. The concept of very large numbers is very interesting to most pupils so use it to help them understand why we need to use alternative ways of representing and manipulating numbers.

Common misconceptions and remediation

- As with all use of powers, pupils tend to confuse 10^n with $10 \times n$. Provide pupils with plenty of opportunity to compare the two and to grasp why they are different. Comparing the two graphically could also help to reinforce the difference,

Probing questions

- Why do 25 ÷ 10 and 250 ÷ 100 give the same answer?
- Why do 6 × 100 and 60 × 10 give the same answer?

Part 1

- This starter is concerned with adding and subtracting 0.1, 0.01 and 0.001 from other decimals and whole numbers. For this activity, ask pupils to write on whiteboards.
- Give pupils an example such as: 'What number is 0.01 more than 6.03?' (= 6.04)
- Now give more examples and ask pupils to hold up the answers on their whiteboards.
- Suggested questions are:
  - What is 0.001 more than 2.008? (= 2.009)
    What is 0.01 less than 5? (= 4.99)
    What is 0.002 less than 3? (= 2.998)
    What needs to be added to 3.234 to make 3.237? (= 0.003)
    What needs to be subtracted from 4.002 to make 3.997? (= 0.005)

Part 2

- Ask the class for a number, for example: 63.
- Produce a spider diagram on the board to show what happens if the number is multiplied or divided by: 10, 10^2, 10^3. Pupils should know the answers to this, but may need to be reminded of the rules about moving digits.
- Ask for the connection between the power and the number of places the digits are moved.
- Ask pupils to complete this pattern:

  $10\,000 = 10^4$
  $1000 = 10^3$
  $100 = 10^2$
  $10 = 10^{\cdots}$
  $1 = 10^{\cdots}$

- Ask how the next number down on the left is obtained from the previous line (divide by 10).
- Ask how the pattern of powers on the right-hand side is obtained (powers decrease by 1).
- Ask pupils to write down the next two lines of the pattern: $= 10^{-1}$; $= 10^{-2}$
- Establish that $\frac{1}{10} = 10^{-1}$ and that $\frac{1}{100} = 10^{-2}$. Give pupils some problems, for example: 7 × 0.1 (= 0.7); 0.4 × 0.1 (= 0.04); 7 ÷ 0.1 (= 70); 4 ÷ 0.01 (= 400). Recall the rules for these or explain how to do them mentally. Repeat with other examples.

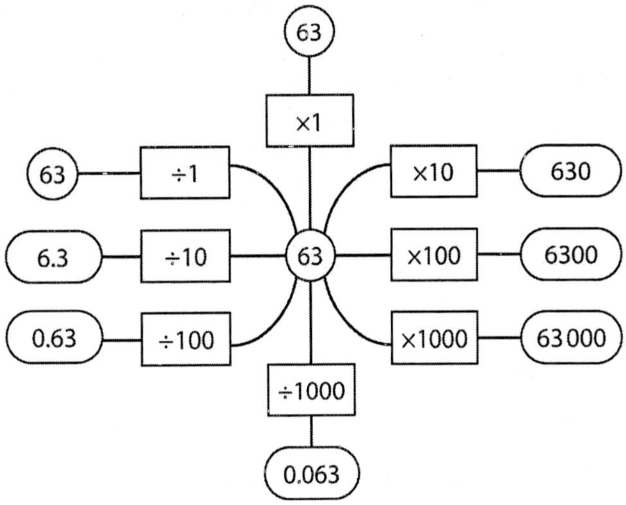

- Use numbers from the target board above to check that pupils can combine the two activities, for example, multiply 4.562 by 10 and round the answer to one decimal place.
- **Pupils can now do Exercise 8A from Pupil Book 2.2.**

Part 3

- Write a variety of numbers on the board (for example, 32, 8, 0.09, 0.312, 48.9, 4598) and ask pupils to multiply and/or divide them by: 10, 102, 0.1, 0.01.
- Discuss the techniques used.

Answers

Exercise 8A

1	**a** 530	**b** 79	**c** 2400	**d** 506.3	**e** 0.3	
2	**a** 0.83	**b** 0.041	**c** 4.57	**d** 0.0604	**e** 347.81	
3	**a** 6430	**b** 685	**c** 35 200	**d** 8074	**e** 2.1	
4	**a** 0.941	**b** 0.00523	**c** 0.568	**d** 0.000 715	**e** 45.892	
5	**a** 31	**b** 678	**c** 560	**d** 0.034	**e** 8.23	**f** 0.009 06
	g 5789	**h** 0.6878	**i** 38	**j** 0.0037	**k** 500	**l** 0.005 43
6	**a** 4250	**b** 567	**c** 451	**d** 0.023	**e** 7.12	**f** 0.008 05
	g 4670	**h** 0.689	**i** 27	**j** 0.000 0049	**k** 600	**l** 0.004 32

7 **a** 0.000 000 000 000 000 000 000 911 g
b 0.000 000 000 000 000 000 911 g

8 **a** 10 000 000 000 000 000 000 000 000 000
b 600 000 000 000 000 000 000 000 000

9 **a i** 300 000 **ii** 30 000 **b** 417

Investigation: Multiplying 9109

A 9109, 18 218, 27 327, 36 436, 45 545, 54 654, 63 763, 72 872, 81 981

B Apart from the first answer, you get the first two digits repeated at the end with the middle number the number you have multiplied by

Lesson 8.2 Large numbers and rounding

Learning objective

- To round large numbers

Resources and homework

- Pupil Book 2.2, pages 139–141
- Intervention Workbook 3, pages 9–10
- Homework Book 2, section 8.2
- Online homework 8.2, questions 1–10

Links to other subjects

- **Science** – to present results to an appropriate degree of accuracy
- **Design and technology** – to give measurements to the degree of accuracy appropriate to the situation

Key words

- approximate
- estimate
- round

Problem solving and reasoning help

- The **PS** question 8 and **MR** question 9 in Exercise 8B of the Pupil Book require pupils to apply their understanding of large numbers and rounding to real-life situations. You could introduce these by asking pupils for their own examples of when they think they might have to use very large numbers or when they might need to round numbers in real life.

Common misconceptions and remediation

- Pupils sometimes have problems with numbers that end in 9, especially if there are several 9s. Pupils may also struggle with numbers with trailing zeros. Provide plenty of opportunity to discuss examples. This applies to large numbers in the same way as it does to smaller numbers. Help pupils to see that in fact making the numbers larger does not make the process of rounding any different.

Probing questions

- How do you go about rounding a number to one decimal place?
- Explain why the rules are the same when you round a large number or a small number.
- Why might it not be possible to identify the first three places in a long-jump competition if measurements are taken in metres, to one decimal place?

Part 1

- Use a target board, as shown. Recall strategies for rounding to one or two decimal places.
- Point to a number and ask pupils to round it to one or two decimal places, as appropriate.

4.562	2.375	1.071	3.222	0.541
0.082	1.629	14.635	3.999	4.814
3.421	8.525	3.688	9.002	1.035
6.455	1.459	1.291	5.927	2.716

Part 2

- In the previous lesson we dealt with powers of 10 such as: 10^2, 10^4 and 10^{-1}.
- Ask pupils if they can 'name' 10^2 and write it as a multiplication problem. They should come up with 'one hundred' and $10^2 = 10 \times 10$.
- Similarly, name and expand 10^3 (one thousand), 10^4 (ten thousand), 10^5 (hundred thousand), 10^6 (million) and 10^9 (one billion). It may be worth pointing out that this is the internationally accepted version of a billion, which was a million millions (10^{12}) in the UK.

- Ask pupils if they can think of any other way we use these numbers. They should come up with column headings in place value.
- Copy the table on the board.
- Ask pupils to read the number (three million, seven hundred and eight thousand, four hundred and thirty-two).

10^6	10^5	10^4	10^3	10^2	10	1
3	7	0	8	4	3	2

- Emphasise that large numbers are read as 'so many millions', 'so many thousands' and finally the last three digits. This is also why large numbers are written in blocks of three, or sometimes with commas between each group of three. However, commas may cause confusion because of the European practice of using a comma as a decimal point.
- Add other large numbers such as: 1 023 708 and 12 007 009, to the table on the board.
- Also give some large numbers in words and ask pupils to write them down, for example:
  - five million, seventy-eight thousand, three hundred and six (5 078 306)
    two million, nine thousand and sixty three (2 009 063).
- **Pupils can now do Exercise 8B from Pupil Book 2.2.**

Part 3

- Write a variety of large numbers on the board, for example:
  4 502 611   5 560 097   2 110 009   7 899 911
- Ask pupils to read out the numbers and round them to the nearest ten thousand, hundred thousand and million.

Answers

Exercise 8B

1 Spain 40 million, Germany 75 million, Italy 57 million, France 56 million, Ireland 4 million, Denmark 6 million

2 **a** 3 550 000 **b** 9 720 000 **c** 3 050 000 **d** 15 700 000

3 **a** 4 700 000 **b** 8 600 000 **c** 4 200 000 **d** 26 800 000

4 **a** 2 million **b** 7 million **c** 3 million **d** 37 million

5 The government, as the figure rounds to 2 million. However, both are actually incorrect, as they should both round to 2.5 million

6 Highest 6499; lowest 5500

7 Highest 8 499 999; lowest 7 500 000

8 Highest 10 547; lowest 8450

9 Micky gets 7500 and Jenna gets 7499

Investigation: Strange addition

A 123456790123456790 …

B 246913580246913580 …

C 370370370703703 70 …

Lesson 8.3 Significant figures

Learning objective

- To round to one or more significant figures

Resources and homework

- Pupil Book 2.2, pages 141–144
- Homework Book 2, section 8.3
- Online homework 8.3, questions 1–10

Links to other subjects

- **History** – to describe prehistoric time scales

Key words

- significant
- significant figure

Problem solving and reasoning and financial skills help

- **FS** question 9 and **PS** question 10 in Exercise 8C of the Pupil Book encourage pupils to think about the use of significant figures in real-life situations, including financial contexts. It helps to connect the concept of significant figures in terms of the degree of accuracy needed to respond effectively in real-life situations. The investigation at the end of the exercise explores significant figures in a more abstract situation, using this to reinforce place value.

Common misconceptions and remediation

- Pupils tend to confuse rounding to decimal places and significant figures. Provide plenty of opportunity for pupils to compare the two. Pupils also struggle with the role of 0, and when this counts as a significant figure. Give pupils practice and answer any questions that arise.

Probing questions

- How do you go about rounding a number to one or two significant figures?
- Why might it not be possible to identify the first three places in a long jump competition if measurements were taken to one significant figure?

Part 1

- Pupils can work in small groups. Give pupils an OHP film and some OHP pens or a large sheet of paper and some felt-tipped pens.
- Give each group the same five (or four) two-digit numbers, for example: 21, 43, 54, 77 and 12, and then the same five (or four) single-digit numbers such as: 2, 3, 5, 6 and 9.
- In a timed session of about five minutes (NO calculators), ask pupils to write down as many multiplication problems as possible, multiplying a two-digit number by a single-digit number.

×	**21**	**43**	**54**	**77**	**12**
2	42	86	108	154	24
3	63	129	162	231	36
5	105	215	270	385	60
6	126	258	324	462	72
9	189	387	486	693	108

- After the allocated time, stop pupils working and check the answers.
- Allocate a score to see which group is the winning group.

Part 2

- Explain how we round numbers to significant figures. For example, we may say, ‘It's about two miles to school’ or ‘It took me 30 minutes to do my homework’. If the actual values are, for example, 2.56 miles and 38 minutes, we are rounding to one significant figure. Explain that we could also say that we are rounding to the nearest whole number or nearest 10.
- However, we can generalise significant figures to any number:
  - The digit with the highest place value is the most significant figure.

- The digit with the next highest place value is the next significant, and so on.

- If you round a number to one significant figure (1 sf), it means that you round it to its highest place value, if you round it to two significant figures (2 sf) you round it to its highest and next highest place value, and so on. For example, in the number 463 521:
  - 5 has the highest place value, so it is the most significant figure. If you rounded the number to 1 sf it would be 500 000.
  - 3 is the next most significant figure; rounding the number to 2 sf would make it 460 000.
- To round significant figures, use the same rules as you did when rounding to powers of 10 in Lesson 8.1. If the digit to the right of the significant figure is less than 5, replace it with zero but leave the digit on the left unchanged. If the digit is 5 or more, replace it with zero and add 1 to the digit on the left. This part is very important as it can cause problems for pupils. Stress that a zero may or may not count as a significant figure depending on its position.
- Note that when the zero is used at the end of a whole number we do not count it as significant. For example: 600 has 1 sf; 720 has 2 sf.
- When the number is less than 1, we start to count significant figures at the first non-zero figure. Any zeros that follow that figure, however, count as significant, for example: 0.007 has 1 significant figure; 0.60 has 2 significant figures.
- When zeros are between other digits, we count them as significant. For example: 6.07 has 3 significant figures; 0.07010 has 4 significant figures.
- **Pupils can now do Exercise 8C from Pupil Book 2.2.**

Part 3

- In pairs or small groups, ask pupils prepare a response to this question: Is there a difference in the way you would deal with the zeros in very large numbers and very small numbers?
- Select group/s to present to the class. The class can provide formative feedback based on an agreed set of criteria such as: Is the explanation clear? Does the group use appropriate examples? Does the group allow time for and/or respond effectively to questions?

Answers

Exercise 8C

1	**a** 4	**b** 2	**c** 3	**d** 3	**e** 1	
2	**a** 300	**b** 4000	**c** 60	**d** 0.9	**e** 0.09	**f** 300
	g 70	**h** 4000	**i** 0.7	**j** 10	**k** 600	**l** 0.08
3	**a** 5300	**b** 50	**c** 9.1	**d** 750	**e** 0.083	**f** 740
	g 350	**h** 6500	**i** 4900	**j** 94	**k** 0.87	**l** 0.0056
4	**a** 2150	**b** 9.56	**c** 0.266	**d** 0.004 10	**e** 0.000 717	**f** 25.9
	g 268	**h** 10900	**i** 0.0827	**j** 1.88	**k** 3 150 000	**l** 254 000
5	**a** 3.2	**b** 8	**c** 14	**d** 17.2	**e** 30	**f** 37
	g 0.064	**h** 0.008	**i** 83.7	**j** 1.0	**k** 20	**l** 23.15
6	**a** 0.33	**b** 0.1429	**c** 0.0769	**d** 0.0909	**e** 1.6	**f** 0.923
	g 2.429	**h** 1.91	**i** 3.92	**j** 3.1	**k** 5.84	**l** 3.364

7 11

8 236

9 £1313, £5429, £4633, £3039, £3817

10 10 000

Investigation: Patterns in calculations

A 0.142 8571, 0.285 7142, 0.428 5714, etc.

B The same numbers 1428571 are used each time and in that cyclic order but different starting positions

C 0.428571428571428571 43 **D** 5

Lesson 8.4 Standard form with large numbers

Learning objective

- To write a large number in standard form

Resources and homework

- Pupil Book 2.2, pages 144–146
- Homework Book 2, section 8.4
- Online homework 8.4, questions 1–10

Links to other subjects

- **Science** – to use and manipulate scientific formulae

Key word

- standard form

Problem solving and reasoning and financial skills help

- **PS** questions 8 and 9 in Exercise 8D of the Pupil Book require pupils to apply their new understanding of large numbers to a familiar context, which they may not immediately associate with very large numbers. Ask pupils to think of everyday examples of questions they could ask that would involve the use of very large numbers, and therefore benefit from the use of SI form. This would be a good time to demonstrate problems on the STEM Nrich website to extend pupils' experience of very large numbers: **http://nrich.maths.org/7278**.

Common misconceptions and remediation

- Pupils need to be confident with the definition of standard form as being a number that is written as $A \times 10n$. It is most important for pupils to grasp that A is always between 1 and 10 no matter if the number is large or small. Be careful how you explain this, as explanations involving moving the decimal point can be misleading and can lead to misconceptions about place value in general. Provide plenty of practice with converting different types of numbers.

Probing questions

- What are the key conventions when using standard form?
- How do you go about expressing a very large number in standard form?

Part 1

- In pairs, encourage pupils to discuss their response to this question: What is the same, or different, about these numbers: 565.3020 and 565.360 50?
- Take feedback, encouraging pupils to draw on their work so far in this chapter. Encourage them to use as much of the language from the previous lessons and compare different ways of rounding and different mathematical language for expressing these numbers.
- Encourage **more able** pupils to come up with examples and design appropriate questions.

Part 2

- Remind pupils about Lessons 8.1 (powers of 10) and 8.2 (large numbers and rounding). Say that they are going to combine these ideas to help them deal with very large numbers.
- Explain to pupils that a different type of notation called standard form is often used to enable us to handle very large numbers effectively.
- Write a very large number on the board, for example: 560 000 000 000 000 000 000 000.
- Ask pupils to describe this number in words. Then ask them to imagine multiplying this number by 1 000 000 and writing this down; then multiplying it by another million, and so on.
- Pupils should start to realise that it will become very difficult to write down the number, let alone describe it in words. Say that this is where standard form comes in.

- Now work through the explanation in the Pupil Book on page 144. Allow pupils to ask questions, and then, as a group, to work through Example 6 on page 145 of the Pupil Book.
- Pupils can then work through Example 7 in pairs and ask any outstanding questions.
- Use grouping and additional support to support or challenge pupils.
- **Pupils can now do Exercise 8D from Pupil Book 2.2.**

Part 3

- Ask pupils: 'How do you know that a very small number might be expressed in standard form?'
- Share ideas and come up with a group explanation.
- Encourage **less able** pupils to use the internet to search for, for example, 'Bitesize' or 'standard form KS3' to help them to explore this question.

Answers

Exercise 8D

1	**a** 5.69×10^3	**b** 1.2×10^6	**c** 9.38×10^5	**d** 7.78×10^4
	e 3.965×10^8	**f** 5.61×10^2	**g** 7.3×10^1	**h** 4.3×10^9
2	**a** 3.4×10^6	**b** 5.6×10^3	**c** 2.6×10^7	**d** 4.5×10^4
	e 2.58×10^8	**f** 5.47×10^5	**g** 2.0×10^8	**h** 5.0×10^5
3	**a** 8.0×10^9	**b** 1.2×10^{10}	**c** 1.5×10^{11}	**d** 1.0×10^{11}
	e 6.7×10^9	**f** 1.55×10^{10}	**g** 1.0×10^{12}	**h** 1.0×10^{18}
4	**a** 2 300 000	**b** 456	**c** 675 000	**d** 3590
	e 9 000 000	**f** 2 010 000	**g** 34 780	**h** 87 300 000
	i 670 000	**j** 38 500 000 000	**k** 780 000 000	**l** 5 390 000 000
5	**a** 2.5×10^5	**b** 1.764×10^7	**c** 1.369×10^5	**d** 8.1×10^7
	e 4.225×10^5	**f** 9×10^8	**g** 2.5×10^{13}	**h** 2.25×10^{12}
6	**a** 7.3×10^7	**b** 2.56×10^4	**c** 7.7×10^6	**d** 2.59×10^5
	e 9×10^8	**f** 7.01×10^7	**g** 3.478×10^7	**h** 1.873×10^{10}
	i 7×10^4	**j** 8.5×10^9	**k** 8×10^6	**l** 8.6×10^7
7	3×10^8			
8	1.09×10^{30}			

Activity: Astronomical numbers

Answers will vary depending on measurement found.

Lesson 8.5 Multiplying with numbers in standard form

Learning objective

- To multiply with numbers in standard form

Resources and homework

- Pupil Book 2.2, pages 147–149
- Homework Book 2, section 8.5
- Online homework 8.5, questions 1–10

Links to other subjects

- **Science** – to use and manipulate scientific formulae

Key words

- No new key words for this topic

Problem solving and reasoning and financial skills help

- This lesson encourages pupils to apply their understanding to real-life concepts with differing degrees of familiarity. Pupils may be surprised at some of the numbers they generate. If so, encourage them to research other examples. The challenge activity at the end of Exercise 8E in the Pupil Book encourages pupils to do some independent research and to work out some very large numbers.

Common misconceptions and remediation

- Pupils need to be confident with the ideas in Lesson 8.1 and in the previous lesson before tackling this lesson. Encourage pupils to revisit those lessons if necessary.

Probing questions

- 'Multiplying makes numbers bigger.' When is this statement true? When is it false?
- 'Division makes numbers bigger smaller.' When is this statement true? When is it false?

Part 1

- Draw a spider diagram on the board similar to the one that is in Lesson 8.1. This time, put the multipliers as powers of 10 and ask pupils to write the answers in standard form.
- Check pupils' understanding before moving on. You could give **less able** pupils partially completed diagrams. The challenge activity could also be done as a collective memory.

Part 2

- Ask pupils to multiply 2000 by 3000. Ask what answer they get. (6 000 000)
- Now ask pupils to write 2000 and 3000 in SI form. (2×10^3), (3×10^3)
- Then ask pupils to write the answer in SI form. (6×10^6)
- Ask pupils to speculate about any patterns. They should say that you multiply the digits, then add the powers to write the answer in SI form. Work through the examples in the Pupil Book.
- **Pupils can now do Exercise 8E from Pupil Book 2.2.**

Part 3

- Give pupils a large sheet of paper. Working in groups, ask them to produce a summary of everything they have done in this chapter.
- Encourage **more able** pupils to give examples that demonstrate the similarities and differences between different representations.

Answers

Exercise 8E

1 **a** 8×10^5 **b** 1.2×10^8 **c** 8×10^7
d 9×10^{13} **e** 3.2×10^{18} **f** 4.2×10^{14}
g 2.1×10^7 **h** 1.0×10^{10} **i** 5.6×10^{11}
j 2.25×10^{10} **k** 1.12×10^{12} **l** 3.6×10^7

2 **a** 9.46×10^9 **b** 1.152×10^8 **c** 1.288×10^{10} **d** 5.51×10^8 **e** 4.672×10^9
f 1.674×10^{11} **g** 2.99×10^6 **h** 1.311×10^{17} **i** 1.296×10^5 **j** 6.561×10^{11}

3 **a** 9.82×10^{10} **b** 7.28×10^7 **c** 7.27×10^9 **d** 2.35×10^8 **e** 4.05×10^{13}
f 5.84×10^{10} **g** 2.95×10^{15} **h** 1.56×10^{18} **i** 6.13×10^{13} **j** 6.42×10^8

4 2.56×10^{11}

5 6.6×10^6 litres

Challenge: Mega-memory

A

1×10^3	KB
1×10^6	MB
1×10^9	GB
1×10^{12}	TB
1×10^{15}	PB
1×10^{18}	EB
1×10^{21}	ZB
1×10^{24}	YB

B 1×10^{18} (an exabyte)

C 5×10^{30}

Review questions

(Pupil Book pages 150–151)

- The review questions will help to determine pupils' abilities with regard to the material within Chapter 8.
- These questions also draw on the maths covered in earlier chapters of the book to encourage pupils to make links between different topics.
- The answers are on the next page.

Challenge – Space – to see where no one has seen before

(Pupil Book pages 152–153)

- This activity is designed to combine the skills covered across this chapter to explore an interesting real-life problem set in a slightly less familiar context. The context is one of the best examples of the use of very large numbers.
- All the information required to answer questions 1 to 6 is provided in the text.
- Questions 7 to 9 require pupils to combine different pieces of information and use skills from other areas of mathematics, in this case formulae and substitution.
- Ask pupils further questions as a warm-up to start working on the questions in the Pupil Book. Here are some examples:
    - What do you know about the universe?
    - What questions might you like to ask about the universe?
- Pupils can now work on the questions individually or in groups.
- Pupils can develop this topic further by using the internet to research more about the questions provided and some of their own questions. Encourage groups to present their findings to the class if there is time.

Answers to Review questions

1 1 m^3 or 1 000 000 000 mm^3
2 £21 000
3 For example 67 000 fans which is 70 000 to 1 sf
4 A hundred million, ten billion, a trillion
5 **a i** 43.5% **ii** 4.35×10^6 **iii** 4.36 **b** 899 999, 5 million, 2250^2, 4.3×10^7
6 1.6×10^{12}
7 **a** 145 000 and 154 999 **b** 15 000
8 **a** 3.99×10^{13} km **b** $114\,000 \times 365 \times 24 \times 40\,000 = 3.99 \times 10^{13}$
9 Take 1.5, which rounds to 2, a difference of 0.5, which is one third of 1.5

Answers to Challenge – Space – to see where no one has seen before

1 9.45×10^{12} km
2 1.3×10^{23}
3 Both 2×10^{11}, speed of light from question 1 is 3×10^8
4 1.52×10^{11}
5 1.64×10^{11}
6 2.5×10^{22}
7 2144 cm^3
8 268 cm^3
9 3.7×10^{80}

9 Interpreting data

Learning objectives

- How to interpret pie charts by the angle size of each sector
- How to use the scaling method to construct pie charts
- How to use scatter graphs
- How to construct scatter graphs

Prior knowledge

- How to interpret data from tables, graphs and charts
- How to read pie charts where the sectors are percentages or simple fractions
- How to find mode, median, mean and range for small data sets

Context

- This chapter builds on previously learnt statistical principles. It extends pupils' use of data and knowledge of how to interpret statistical diagrams and charts. Pupils will study the way that different types of graphs and charts are used – a vital aspect if pupils are to understand and question or analyse the data being presented. The video clip at this link is an example of the power of statistics: **http://www.gapminder.org/videos/the-river-of-myths/**

Discussion points

- For a given pie chart, make up three questions that you can answer using the information represented.
- What information do you need to work out the total in each category of a pie chart?
- How do you know if there is no correlation in a scatter graph?

Associated Collins ICT resources

- Chapter 9 interactive activities on Collins Connect online platform
- *Constructing pie charts* video on Collins Connect online platform

Curriculum references

Reason mathematically

- Explore what can and cannot be inferred in statistical and probabilistic settings, and begin to express their arguments formally.

Statistics

- Describe, interpret and compare observed distributions of a single variable through: appropriate graphical representation involving discrete, continuous and grouped data; and appropriate measures of central tendency (mean, mode, median) and spread (range, consideration of outliers)
- Construct and interpret appropriate tables, charts, and diagrams, including frequency tables, bar charts, pie charts, and pictograms for categorical data, and vertical line (or bar) charts for ungrouped and grouped numerical data

- Describe simple mathematical relationships between two variables (bivariate data) in observational and experimental contexts and illustrate using scatter graphs.

Fast-track for classes following a 2-year scheme of work

- You could combine Lesson 9.1 and Lesson 9.2. Much of the material in Lesson 9.3 and Lesson 9.4 will be new to the class. However, these two lessons could also be combined. Make certain that pupils have a good grasp of correlation before moving on.

Lesson 9.1 Pie charts

Learning objective

- To work out the size of sectors in pie charts by their angles at the centre

Resources and homework

- Pupil Book 2.2, pages 155–159
- Intervention Workbook 3, pages 69–72
- Homework Book 2, section 9.1
- Online homework 9.1, questions 1–10

Links to other subjects

- **Geography** – to interpret weather and climate charts
- **Physical education** – to interpret data on sports participation

Key words

- pie chart
- sector

Problem solving and reasoning help

- **PS** questions 8 and 9 in Exercise 9A of the Pupil Book require pupils to apply their learning to real-life problems involving pie charts. If **less able** pupils struggle, model a couple of examples with them. The activity at the end of the exercise requires pupils to apply their understanding in a more general way, similar to the way you might respond in real life. Pupils are also required to make links to fractions. You may want to revisit comparing simple fractions if **less able** pupils struggle with the required comparisons.

Common misconceptions and remediation

- Pupils struggle to transfer their understanding of angles, fractions and percentages to interpreting pie charts. Stress these links with careful use of correct mathematical language.

Probing questions

- Show pupils a pie chart and ask them to make up a statement or question for the pie chart using one or more of these words: fraction, percentage, proportion.

Part 1

- On the board, draw a rectangle and divide it into four equal parts.
- Shade one part and ask the class '*What fraction is shaded?*' ($\frac{1}{4}$)
- Shade two parts and ask the class '*What fraction is shaded?*' ($\frac{1}{2}$)
- Draw another rectangle on the board and divide it into six equal parts.
- Shade one part and ask the class '*What fraction is shaded?*' ($\frac{1}{6}$)
- Shade two parts and ask the class '*What fraction is shaded?*' ($\frac{1}{3}$)
- Write 60 on the board and ask: '*What is half of 60?*' (30); '*What is* $\frac{1}{3}$ *of 60?*' (20); '*What is* $\frac{1}{4}$ *of 60?*' (15); '*What is* $\frac{1}{6}$ *of 60?*' (10)
- Discuss with pupils how they worked these out – by splitting 60, or by dividing.
- Repeat the above for 120, then for 300.

Part 2

- On the board, draw a circle. Tell pupils that they will look at ways in which a circle can represent data. A circle represents a pie.

- Draw some simple radii in the circle and explain that each category is represented by a sector of the circle (a slice of the pie).
- Draw another circle (pie chart) with a 90 degree angle. Explain that this pie chart represents the car sales in a showroom on one weekend.
- Label the 90-degree sector 'British', saying that, 'This sector represents British cars sold.'
- Label the rest 'Foreign', saying, 'This sector represents foreign cars sold.'
- Ask, 'If 40 cars were sold, how many of these sales were British cars?'
- Talk through with the class that the sector shown is one quarter of the pie, so $\frac{1}{4}$ of the 40 cars that were sold were British. Ask: 'What is $\frac{1}{4}$ of 40? (10)
- Then ask: 'How many foreign cars were sold?' (The rest = 30)
- Draw another pie chart (as per Example 2 on page 156 of the Pupil Book) on the board. Say that the pie chart represents how one country dealt with dangerous waste in 2013.
- Ask, How much dangerous waste was dealt with by:
  a being landfilled **b** being burnt **c** dumping in sea **d** chemical treatment.
- From the pie chart pupils should see that:
  a $\frac{1}{2}$ of the waste was landfilled = $\frac{1}{2}$ of 3000 kg = 1500 kg landfilled
  b $\frac{1}{4}$ of the waste was burnt = $\frac{1}{4}$ of 3000 kg = 750 kg burnt
  c $\frac{30}{360} = \frac{1}{12}$ of the waste was dumped at sea = 3000 ÷ 12 = 250 kg dumped at sea
  d $\frac{60}{360} = \frac{1}{6}$ of the waste was treated by chemicals = 3000 ÷ 6 = 500 kg treated by chemicals.
- **Pupils can now do Exercise 9A from Pupil Book 2.2.**

Part 3

- Explain that both pie charts they looked at were simple to interpret, with easy fractions. Say that though this is not always the case, we can also estimate numbers by using percentages.
- Draw a pie chart with random angles. Label the sectors with the favourite colours such as red, green and blue of 100 pupils. Ask pupils to estimate the percentage of each sector, or the number of pupils who chose each colour as their favourite.

Answers

Exercise 9A

1 **a** 225 **b** 150 **c** 525
2 **a** 75 **b** 75 **c** 50 **d** 100
3 **a** 45 **b** 45 **c** 60 **d** 90 **e** 75 **f** 135 **g** 90
4 **a** 10 **b** 5 **c** 15
5 **a** 6 hrs **b** 6 hrs **c** 4 hrs **d** 8 hrs
6 **a** 60 **b** 20 **c** 100 **d** 30 **e** 30
7 **a** 45 **b** 72 **c** 45 **d** 54
8 Y7-560, Y8-420, Y9-280, Y10-210, Y11-210
9 Northern Ireland - 19, Wales - 93, Scotland - 100, England - 420

Activity: Population in a pie chart

A 80+
B 0–19 … 1 247 000, 20–39 … 1 546 280, 40–59 … 1 197 120, 60–79 … 847 960, 80+ … 149 640

Lesson 9.2 Creating pie charts

Learning objective

- To use a scaling method to draw pie charts

Resources and homework

- Pupil Book 2.2, pages 159–161
- Homework Book 2, section 9.2
- Online homework 9.2, questions 1–10

Links to other subjects

- **History** – to compare historical data using pie charts
- **Geography** – to interpret weather and climate charts
- **Physical education** – to interpret data on sports participation

Key words

- frequency
- frequency table
- scaling
- total frequency

Problem solving and reasoning help

- **PS** questions 5 and 6 of Exercise 9B in the Pupil Book encourage pupils to apply their learning in order to deepen their understanding of pie charts. To help **less able** pupils, model several examples with them.

Common misconceptions and remediation

- Pupils often forget to change frequencies to angles. Another problem is drawing sector angles accurately. **Less able** pupils may need practice with measuring and drawing angles with a protractor.

Probing questions

- When drawing a pie chart what information is required in order to calculate the angles?
- What are the main advantages of presenting information using a pie chart?
- In which jobs might you be asked to present information using a pie chart?

Part 1

- Draw unlabelled angles of 30°, 60° and 120° on the board.
- Tell pupils that there is a pattern to the sizes of the angles. Ask them to estimate the size of the angles. Now say that the pattern is 'doubling', and let them change their estimates if they wish. Ask for and then write pupils' answers on the board. Comment on their accuracy.
- Now draw three bars or vertical lines from the same base level, with heights in the ratio of: 3 : 4 : 5. Say that the height of the first bar represents 15 people, and that, as before, there is a pattern to the heights. Ask the class to estimate the heights represented by the other bars.
- Now ask them to think of angles, and that you need three angles that add up to 360°. They need to represent 12 people altogether. Ask the class to work out what angle is needed for four people, then three people and finally five people.
- Ask them to explain how they worked it out. Check to see if anyone did it differently.
- Now say that you want three facts about a pie chart (for example, angles add up to 360°, need labels or a key, circular).

Part 2

- Explain that in this lesson pupils will construct pie charts from data that have already been collected, called secondary data. Point out that data that they collect is called primary data.

- Draw a circle on the board or OHP and say that this represents everyone in the room. Ask pupils how they would show on the pie chart that half the class are boys and half are girls. Then ask how you would show that one-quarter wear glasses and the rest do not.
- Write 36 on the board. Say: 'A pie chart should show 36 people. How many degrees per person?' (10). Continue this, using factors of 360, changing the number of people each time.
- Ask pupils to give you numbers that add up to 36. Write them on the board in a table as shown in the example. Repeat that there are 10 degrees per person. Ask pupils to tell you how big each angle needs to be on the pie chart.

Number of people	Angle
20	200
7	70
6	60
3	30

- Show pupils how you want them to set out their working for pie charts, either in a table or as in Example 3 on page 160 of the Pupil Book, where the complete calculation is written out each time.
- Ask pupils to copy the above example into their exercise books.
- Explain to pupils that for a pie chart they can work the question backwards by using the fraction of 360 to calculate how many items each sector represents. Use the example provided on the board to explain this clearly, such as for the smallest angle (30).
- **Pupils can now do Exercise 9B from Pupil Book 2.2.**

Part 3

- Explain to the class that, to make the work easier, the pie charts they have been doing all used numbers that are factors of 360. Add that this will not always be the case and that they may need to use calculators to do some of the work. This may mean that they will need to round off the angles to the nearest degree.

Answers

Exercise 9B

1 Angles of 120°, 70°, 80°, 40°, 50°
2 Angles of 99°, 72°, 54°, 81°, 18°, 36°
3 Angles of 45°, 60°, 105°, 75°, 60°, 15°
4 Angles of 102°, 48°, 126°, 18°, 66°
5 **a** The angles for 4 and 5 rooms would be far too small
b 1 room – 112°, 2 rooms – 233°, 3+ rooms- 15°
c Check pupils' pie charts
6 Angles of 14°, 18°, 40°, 133°, 155°

Challenge: World energy consumption

A Oil 38, hydro 8, nuclear 5, coal 30, gas 20
B Estimating
C Angles of 137°, 29°, 18°, 108°, 72°

Lesson 9.3 Scatter graphs and correlation

Learning objective

- To read scatter graphs
- To understand correlation

Resources and homework

- Pupil Book 2.2, pages 162–165
- Homework Book 2, section 9.3
- Online homework 9.3–4, questions 1–10

Links to other subjects

- **Geography** – to make data comparisons for different countries
- **Physical Education** – to compare records for different athletes

Key words

- correlation
- scatter graph

Problem solving and reasoning help

- The **MR** and **PS** questions in Exercise 9C test pupils' understanding of correlation. Most pupils should be able to state the nature of correlation from the scatter graph patterns without too many problems. These questions, however, explore the underlying reasons for the correlation. For example, **MR** question 5 b is positive correlation but the reason this is positive is because the greater the weight of the parcel the greater the cost to post it. Make certain that pupils have this deeper understanding of correlation before moving on.

Common misconceptions and remediation

- Explain to pupils that correlation does not imply connection, which is a common problem area in the interpretation of correlation.

Probing questions

- Describe the type of correlation you would expect from each of these sets of data if you were to plot them on a scatter diagram:
  - height and weight
  - eye colour and IQ
  - hours watching TV and time spent on homework.

Favourite colour	Class 1	Class 2
Blue	9	15
Red	8	5
Yellow	5	6
Other	8	4

Part 1

- Ask pupils to give facts about the table. For example, blue is the favourite in both classes, three times as many like blue as red in Class 2, twice as many like 'other' in Class 1.
- Change the numbers or the headings and repeat.
- Now ask pupils to draw a blank table the same size and with the same headings, and ask them to try to complete it with these facts (pupils may work in pairs or groups):
  - There are 30 pupils in each class.
  - In Class 1, 25 did not pick red, and no one picked yellow.
  - In Class 2, three times as many picked red as in Class 1.
  - 10 more picked blue than red in Class 1.
  - In Class 2, only three picked 'other' and twice as many picked blue as yellow.
- Now ask how many picked yellow in Class 2. You may need to repeat the statements.

Answer

Favourite Colour	Class 1	Class 2
Blue	15	8
Red	5	15
Yellow	0	4
Other	10	3

Part 2

- Say that now pupils will look at pairs of sets of data to see if there are any connections or relationships between them. Ask what happens to the height of children as they get older.

- Sketch a graph on the board of height against age and plot a cross near the origin. Say that this cross represents a very young child. Ask pupils to tell you where the cross will be one year later. Plot the new cross. Continue with this until you have a few crosses with a positive correlation. Introduce the words ‘correlation’ and ‘positive correlation’. If appropriate, discuss why the crosses do not fall on a perfect straight line and about strong or weak correlation.
- Now sketch a negative correlation graph on the board and ask pupils to suggest what the labels could be, for the axes, for example, ‘Value of a car’ and ‘Age’.
- Discuss the idea of ‘negative correlation’ and then introduce the idea of ‘no correlation’.
- Draw a graph of each type on the board and ask the class what correlation each one has.
- **Pupils can now do Exercise 9C from Pupil Book 2.2.**

Part 3

- Emphasise that there are other types of statistical graphs as well as those already covered. Mention bar charts, frequency diagrams and line graphs.
- Discuss the importance of being able to choose the graph they need to use and justifying their choice. Say that in the next lesson they will look at creating scatter graphs.

Answers

Exercise 9C

1 **a** Positive, warmer weather so more deckchairs needed
b Negative, wet weather so less deckchairs needed
c Positive, more umbrellas sold when it is wet
d Negative, less ice creams sold when it is wet
e Positive, more ice creams sold when it is warm
f None, sale of umbrellas unaffected by temperature

2 **a** Positive, “the taller you are, the larger shoe the shoe size”
b No correlation between weight and shoe size
c Positive, “the higher the mass, the larger the collar size”
d No correlation between height and collar size

3 **a** Positive, the higher the score in test A, the higher the score in test B
b No correlation between maths and English results
c Positive, the higher the score in maths, the higher the score in science
d Negative, the higher the score in English, the lower the score in science

4 **a** Negative, the higher the price, the fewer goals let in
b No correlation between age and goals let in
c Negative, the higher the price, the fewer goals let in
d Negative, the older the goalkeeper, the fewer goals let in

5 **a** No correlation between distance and cost
b Positive, the higher the weight, the greater the cost
c Positive, the longer the distance, the longer the time
d No correlation between mass and time

6 Billy – graph C, Terry – graph A, Suzie – graph B

7

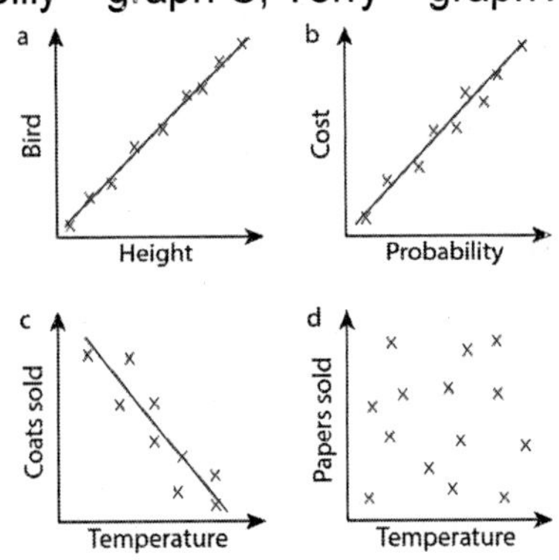

8

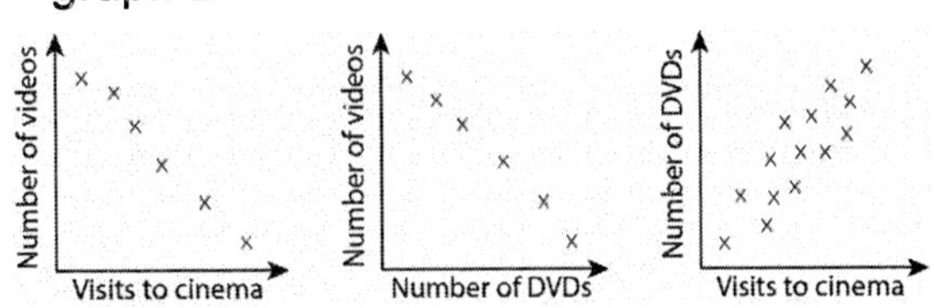

Activity: Correlation in circles

Yes, there is positive correlation between diameter and circumference

Lesson 9.4 Creating scatter graphs

Learning objective

- To create scatter graphs

Resources and homework

- Pupil Book 2.2, pages 166–167
- Homework Book 2, section 9.4
- Online homework 9.3–4, questions 1–10

Links to other subjects

- **Geography** – to make data comparisons for different countries
- **Physical Education** – to compare records for different athletes

Key words

- No new key words for this topic

Problem solving and reasoning help

- **PS** question 4 could be used to introduce the concept of a line of best fit in order to estimate missing results. This could lead onto an explanation of interpolation and extrapolation for **more able** pupils.

Common misconceptions and remediation

- Make sure that pupils understand that scatter diagrams are different to line graphs, as some pupils may try to join the plot points.

Probing questions

- Show pupils a scatter diagram such as the one shown and ask questions such as:
    - What correlation does the graph show?
    - Did the pupil with the highest or lowest score in Test A get the highest score in Test B?
    - What was the biggest difference between pupils' scores in Test A and Test B?
    - What is the range of scores for Test A and Test B?
    - Is this a good choice of graph for indicating in which test pupils did best?

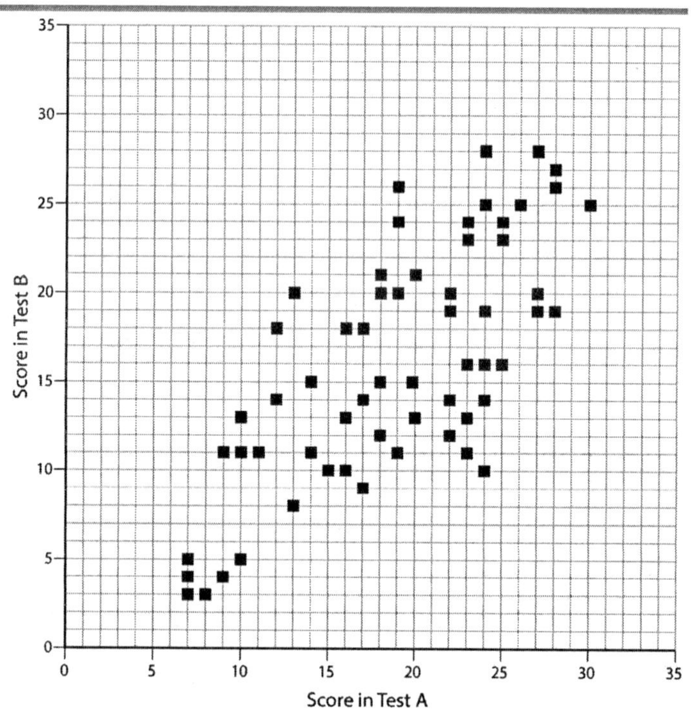

Part 1

- On the board, write: 'Children eat more junk food than adults.' Tell pupils to imagine that they have collected lots of data from children and adults to try to prove or disprove this statement.
- Write some facts on the board. What sort of diagrams could be used to show these facts?
- 75% of the children and 45% of the adults surveyed eat junk food at least once a week.
- The number of times that the pupils surveyed eat chips each week is: 3, 3, 2, 4, 4, ….

- Suppose that a table is given to compare age with the approximate number of times people eat junk food each month. Two sets of data, one for children and one for adults, are obtained about how often they go to burger restaurants.
- You may wish to add to the list. Encourage pupils to give several choices for each.

Part 2

- Review correlation and the different ways to describe it: positive, negative or no correlation.
- Say that these are the three main types, but that some correlations can be seen more clearly than others; some may be obviously positive; others show a slight positive correlation only.
- Ask pupils to estimate how far from school they live in kilometres, and how many minutes it usually takes them to get from home to school. Write these on the board, for example:

MW	3.5 km	20 minutes
AB	1 km	10 minutes

- Draw a pair of axes on the board to illustrate these data, while discussing the labels and the numbers that need to go on the axes. Mark the horizontal time and the vertical distance.
- Plot each pupil on the board with a small cross. (Add pupils' initials to the crosses.)
- Discuss the correlation that the graph shows. Explain that it is important, when plotting scatter graphs, to do so as accurately on graph paper as possible, using sensible scales.
- **Pupils can now do Exercise 9D from Pupil Book 2.2.**

Part 3

- Ask the class to suggest who might use scatter graphs. Some examples are:
    - teachers, to plot school tests and determine what an absent pupil might have scored
    - researchers, trying to find links between factors such as life expectancy and smoking
    - retailers of various commodities such as ice cream or umbrellas.

Answers

Exercise 9D

1 **a** Check pupils' graphs
  b The older the pupil the more money they spend **c** Around 13 years old
2 **a** Check pupils' graphs
  b The more time they spend on homework the less tv they watch **c** 10 hrs
3 **a** Check pupils' graphs
  b The older the car the less it is worth **c** £9000
4 Celia-8 on test B, Ida-50 on test B,
  Les-50 on test A, Ulla-30 on test A

Problem solving: Fish food

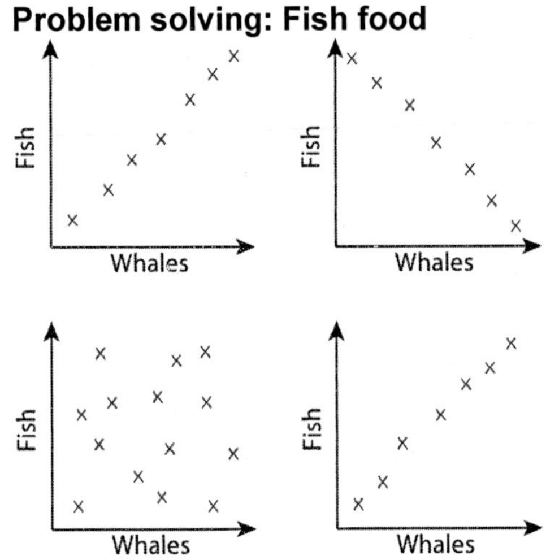

Review questions (Pupil Book pages 168–169)

- The review questions will help to determine pupils' abilities with regard to the material within Chapter 9.
- These questions also draw on the maths covered in earlier chapters of the book to encourage pupils to make links between different topics.

Challenge – Football attendances (Pupil Book pages 170–171)

- Introduce this activity once you have covered the previous work on statistics.
- Ask pupils who they think is the median height of the class (you could revise 'median').
- After some discussion, ask pupils how they can find out. (All line up in order of size and see who is in the middle.) While standing in line, ask pupils if they can see a mode, taking the heights to the nearest centimetre.
- Write four numbers on the board such as 5, 7, 8, 10. What is the median? Pupils may say that there is no middle number. Explain that the median is the middle of the two middle numbers – or halfway between them. Here the median of 5, 7, 8, 10 is: $7\frac{1}{2}$.
- Ask the class for the median of the following numbers if the middle two numbers in an ordered list are: 8, 10 (median 9); 16, 17 (median $16\frac{1}{2}$); 10, 13 (median $11\frac{1}{2}$).
- Continue with other examples that have only a small difference between the two numbers.
- Discuss football attendances. Ask pupils if they know the typical attendance numbers at a large game involving, for example, Manchester United (70 000) and a small crowd at a lower division side (2000–5000)?
- Remind pupils that some cities such as Manchester, Birmingham and London have more than one football team. This is also true of Sheffield, which has two teams: Sheffield Wednesday and Sheffield United. The teams play home games at different weekends and the size of the crowd usually reflects how well the team is playing and at what level.
- Ask pupils to study the data in the table on page 170 of the Pupil Book. Make sure that pupils know what is meant by:
  - Level (1 is the top division, 2 is the next division and so on)
  - Place (the position in the division at the end of the season)
  - Attendance (the average attendance over the whole season).
- The data-handling expected in this spread is:
  - reading from a table
  - averages, mode, median and mean
  - grouped bar chart
  - pie charts.
- Pupils should remember what is meant by mode, median and mean. Ask pupils if the average attendances in the tables were actually the mode, median or mean. Discuss the mean being the most likely. Why? (It is the average that uses every person who attended, allowing for strange distributions.)
- Ask why it is most unlikely to have been made by the mode. (It is very unlikely that more than one attendance in a season will be the same as another.)
- Ask pupils to work through questions 1 to 5 in the activity in the Pupil Book.

Answers to Review questions

1 **a** 36 and 324 **b** Not possible to tell

2 **a** Angles 60°, 180°, 120° **b** 24

3 **a** N **b** True **c** 70

4 **a** Check pupils' graphs **b** Negative correlation **c** Just over 3 minutes

Answers to Challenge – Football attendances

1

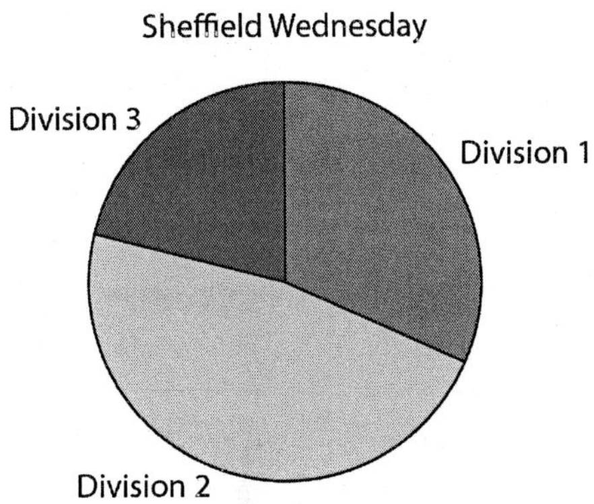

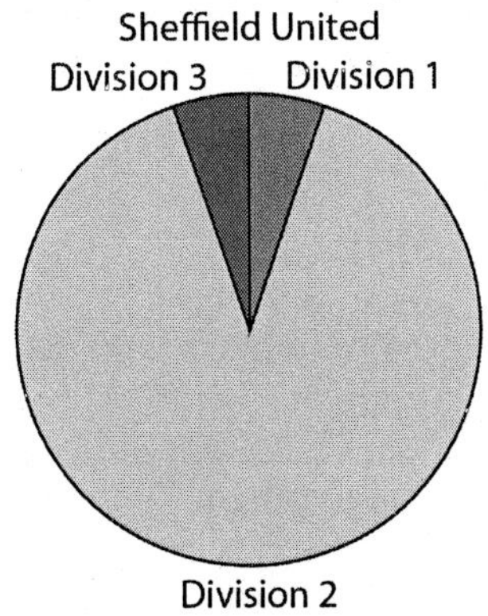

2

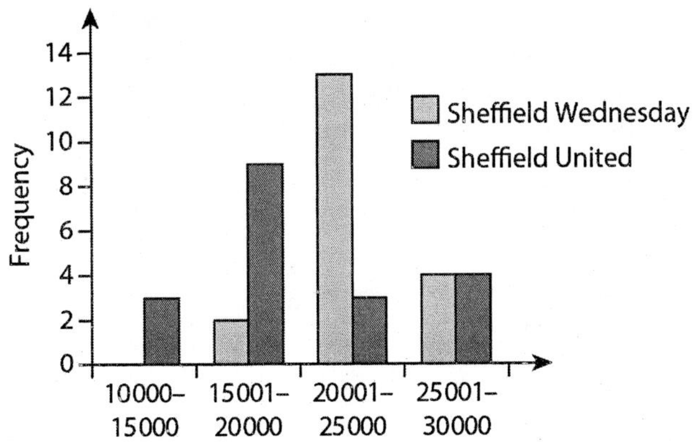

For example, most of the Sheffield Wednesday attendances are higher than 20 000, and most of the Sheffield United attendances are less than 20 000.

3 Sheffield Wednesday = 10 889; Sheffield United = 17608

4 Sheffield Wednesday = 1.9; Sheffield United = 2

5 Check pupils' scattergraphs.

10 Algebra

Learning objectives

- How to write and simplify expressions involving all four operations
- How to simplify expressions that have a number of terms
- How to multiply out brackets in algebraic expressions
- How to identify equivalent expressions
- How to write algebraic expression involving powers

Prior knowledge

- What an algebraic expression is
- How to interpret simple algebraic expressions
- How to write simple algebraic expressions

Context

- Introduce algebra as a universal language with rules that are used all over the world.
- Mathematicians have been developing the rules of algebra for over 3000 years. The Babylonians used a form of algebra when they wrote on clay tablets, some of which have survived until today.
- Discuss a range of examples in which algebra is used. For example, extend on the classic handshakes problem, or compare the Königsberg bridge problem to a modern online community such as Facebook. Search on the internet for 'handshakes problem', 'Königsberg bridge problem' or look at: **http://nrich.maths.org/6708 or http://nrich.maths.org/2484**

Discussion points

- I think of a number and add 12. Do you know what my number is? Why or why not?
- 'I think of a number and add 12. The answer is 17. Do you know what my number is? Why or why not?

Associated Collins ICT resources

- Chapter 10 interactive activities on Collins Connect online platform
- *Human Tower* Wonder of Maths on Collins Connect online platform

Curriculum references

Algebra

- Use and interpret algebraic notation, including:
  - Substitute numerical values into formulae and expressions, including scientific formulae
  - Understand and use the concepts and vocabulary of expressions
- Simplify and manipulate algebraic expressions to maintain equivalence by:
  - collecting like terms

Develop fluency

- Use algebra to generalise the structure of arithmetic, including to formulate mathematical relationships

Solve problems

- Develop their mathematical knowledge, in part through solving problems and evaluating the outcomes, including multi-step problems

Fast-track for classes following a 2-year scheme of work

- Pupils should have met the concepts in Lesson 10.1 and Lesson 10.2 before. Work through a couple of examples to check pupils' understanding and then move on to Lesson 10.3.

Lesson 10.1 Algebraic notation

Learning objective

- To simplify algebraic expressions involving the four basic operations

Resources and homework

- Pupil Book 2.2, pages 173–174
- Intervention Workbook 3, pages 34–36
- Homework Book 2, section 10.1
- Online homework 10.1, questions 1–10

Links to other subjects

- **Science** – to understand scientific formulae

Key word

- algebraic expression

Problem solving and reasoning help

- The challenge at the end of Exercise 10A in the Pupil Book uses the strategy of 'always true' and 'not true' to explore the general rules of algebra and to identify algebra as a generalised form of number. The challenge could be done in small groups or as a class activity, as it is crucial to encourage strong mathematical dialogue among pupils.

Common misconceptions and remediation

- Pupils often struggle to grasp that letter symbols represent numbers or variables rather than labels, and try to substitute particular values for the letters. For example, 5*b* cannot mean five bananas. If the *b* is interpreted as bananas when pupils come across $3b + b + 6$, they will ask: '6 what?' Or, if a stands for apples, then what does *ab* mean? This approach will lead to serious confusion when multiplying brackets. Give pupils plenty of opportunity to reflect on the use of algebra as generalised number and to make clear links to the rules they have learnt. Discuss the power of algebra or generalisation.

Probing questions

- How do you know if a letter symbol represents an unknown or a variable?
- How are these two expressions different?

Part 1

- Ask pupils to do a countdown from 10, 9, 8 ... to zero.
- Now ask pupils to do a countdown from 10, but in steps of $\frac{1}{2}$: 10, $9\frac{1}{2}$, 9, $8\frac{1}{2}$ ... to zero.
- Then ask pupils to do a countdown from 10 in steps of 0.4: 10, 9.6, 9.2, 8.8 ... to zero.
- If this proves to be very difficult mentally, write the numbers on the board as pupils respond.
- With pupils in pairs (or individually), sort out a sequence such as down a class row and back up. Ask pupils to do an individual countdown to zero, for example, from 10 in steps of 0.3.
- You can extend the countdown to start from 100 and use steps of numbers such as 2.5.

Part 2

- Show pupils some well-known symbols such as road signs. Ask what each of them means.
- Ask what is meant by the symbols: +, ×, ÷, – (shorthand for our arithmetic operations).
- Tell pupils that it is a very powerful tool to be able to use a letter to substitute a number, as this has its own shorthand, which is what we need to be able to recognise.
- Explain that another tool in algebra is to leave out the multiplication sign. Ask pupils if they know why (because *x* is widely used in algebra so to use it may cause confusion; it is also easier to write the expression). Show pupils how to write the favoured 'curly' style *x*.

- Introduce these conventions: $5 \times m$ is written as $5m$ (leave out the × sign); $m \times 5$ is written as $5m$; $b \times 2a$ is written as $2ab$ (numbers come to the left of letters); $1x$ is written as x; $a \div b$ is written as $\frac{a}{b}$; $2 \div (a + 1)$ is written as $\frac{2}{a+1}$ (write the division as a fraction).
- Now simplify some expressions involving multiplication and division:
  $2 \times m \times 3 = 2 \times 3 \times m = 6 \times m = 6m$; $3 \times a \times 2b = 3 \times 2 \times a \times b = 6ab$; $\frac{3a}{b} = a$ (for example, one-third of three apples is one apple).
- Now ask pupils to do questions 1 and 2 of Exercise 10A from Pupil Book 2.2.
- Discuss the = sign, and explain that we use it to mean 'has the same value as', and not 'is exactly the same as'. For example, in $3 + 4 = 4 + 3$, each side has the same value, but they are not exactly the same as they are written a different way around.
- Explain that we can use algebra to help us recognise useful, general patterns and properties of our operations in arithmetic. For example, $a + b = b + a$, demonstrates that we can always add two numbers the other way around and get the same answer.
- Ask pupils if the same thing works for subtraction. For example, is $a - b = b - a$? (No)
- Does it work for multiplication? For example, is $ab = ba$? You may need to show a few examples before pupils are convinced, for example, $3 \times 4 = 4 \times 3$. Contrast this with division.
- Explain to pupils that it is very easy to misuse the equals sign, and we must be careful not to do that. Lead the class through solving the equation $3x + 2 = 23$ on the board.
- Write out each stage clearly, emphasising that the same thing must always be done to BOTH sides in order to keep them equal. Show clearly, in steps, how to set out the solution with equals signs in a column, to end up with: $x = 7$.
- **Pupils can now complete Exercise 10A from Pupil Book 2.2.**

Part 3

- Go through the shorthand routines of algebra, and ask pupils why they are useful. Also discuss the misuse of the equals sign, for example: $2x + 3 = 18 = 2x = 15 = x = 7.5$
- Discuss with the class why this is wrong.

Answers

Exercise 10A

1	**a** $4k$	**b** $3t$	**c** $2x$	**d** $20y$	**e** ab	**f** nm	**g** $\frac{1}{2}e$ or $\frac{e}{2}$	**h** $\frac{1}{4}r$ or $\frac{r}{4}$
2	**a** $2ab$	**b** 40	**c** $10e$	**d** $80w$	**e** abc	**f** $3n$	**g** $7.5x$	**h** $\frac{3}{4}bh$
3	**a** a^2	**b** $4x^2$	**c** $9n^2$	**d** $1.4t^2$	**e** $3.2a$	**f** $\frac{1}{5}d^2$	**g** $\frac{1}{2}x^2$	**h** $\frac{2}{3}m^2$
4	**a** $2(n + 1)$		**b** $4(t + 12)$		**c** $8(3 + k)$		**d** $0.5(t - 6.4)$	
	e $4(k + 2)$		**f** $2(12 - y)$		**g** $8(x - 2)$		**h** $6(40 - y)$	
5	**a** $2 + 3a$	**b** $9x$	**c** $7.7w$	**d** $ab - 1.4$	**e** $5 + 4a$	**f** $8.5d$	**g** $12 - 7n$	**h** $13f + 9$
6	**a** $6n$	**b** $20b$	**c** $2d$	**d** $0.5q$	**e** $10k$	**f** $24g$	**g** $12t$	**h** h
7	**a** $4n^2$	**b** $12d^2$	**c** $8p^2$	**d** $3a^2$				

8 **a** Answers given in Pupil Book **b** $\frac{y}{2}$ **c** $\frac{t}{1.5}$ **d** $\frac{24}{n}$

9 **a** Answers given in Pupil Book **b** $\frac{1}{3}m$ **c** $\frac{1}{4}n$ **d** $\frac{1}{2}ab$

10 **a** $3y + 2x$ **b** $5xy$ **c** $6xy$ **d** $x + 5y$

Challenge: Is it true?

A i and iii

B Substitute values of a and b that give different answers

Lesson 10.2 Like terms

Learning objective

- To simplify algebraic expressions by combining like terms

Resources and homework

- Pupil Book 2.2, pages 175–176
- Homework Book 2, section 10.2
- Online homework 10.2, questions 1–10

Links to other subjects

- **Science** – to manipulate scientific formulae

Key words

- term
- like terms

Problem solving and reasoning help

- The investigation at the end of Exercise 10B in the Pupil Book pulls together what pupils have learnt in the first two lessons. It also helps pupils to see how they can use algebra to self-check. Explain that in much of their work on algebra pupils should not get a wrong answer, as in many cases they can use the algebra itself to check their solution.

Common misconceptions and remediation

- Pupils often struggle to identify what constitutes 'like' in this context. Provide opportunities for pupils to compare and discuss what are and what are not like terms. Pupils often do not make the connection from Lesson 10.1 that when we write a letter with no number in front such as x, that there is in fact a 1 in front of the x and we have 'one x'. Say that this is just one of the conventions of writing mathematics that has developed over the years.

Probing questions

- Can you write an expression that simplifies to $3n + 6$?
- How would you explain to someone how to combine like terms?
- What are the important points to remember when combining like terms?

Part 1

- Say that this is about using rules and mental maths (no pen and paper).
- Pupils must give you a number; then you will give them a number. Then pupils must guess the rule. The idea is to see who can guess the rule first.
- Start with the rule: 'Double and add 1'. Ask pupils in turn for a number such as 7, and give them the reply after you have applied the rule (in this case, 15).
- Use other rules, such as: 'Take away from 50'; 'Add 3 and double'.
- Ask some pupils to think of a rule, and others should give these pupils the numbers.
- You could develop this into a team game, with half the class versus the other half. Each team takes a turn to think of the rule and then to guess. The rule must be found within so many turns, or a score must be kept of how many guesses had to be made, or of how many numbers had to be processed.

Part 2

- Ask the class for a quick way of writing 2 + 2 + 2. (3 × 2)
- Replace 2 by a and show that $a + a + a$ can be written more concisely as $3 \times a$ or $3a$.
- Write $3a + 2a$, $3d + e - 2f$ and $\frac{x}{4} + x^2$ on the board. Remind the class of the meanings of 'expression' and 'term'. (An algebraic expression is made up of individual terms.)
- Write the expression $3a + 2a$ on the board and ask for its terms. ($2a$, $3a$)

- Ask how the terms are alike. (Both: contain *a*; are multiples of *a*)
- Demonstrate how like terms can be combined to simplify an expression, for example, $3a + 2a$ can be simplified to $5a$ (use the comparison of adding three apples and two apples).
- Simplify some other expressions containing like terms, for example, $2a + 3a$, $2ab + 3ab$.
- Write the expression $3a + 2b$ and ask if it can be simplified. Describe the terms $3a$ and $2b$ as 'unlike', which means that they cannot be combined, so the expression cannot be simplified. (Use the comparison of adding three apples and two bananas.)
- Go through some more examples of unlike terms:
  $3ab + 2a$ (like terms must have exactly the same letters)
  $3a^2 + 2a$ ($3a^2 = 3 \times a \times a$)
- Simplify some more expressions containing like and unlike terms, for example:
  $4e + 3e + 5f = 7e + 5f$ $4x - x = 3x$ (remind pupils that x can be written $1x$)
  $2b - 7b = -5b$ (because $2 - 7 = -5$) $-a - a = -2a$
  $4x + 2y - 3x = 4x - 3x + 2y = x + 2y$ (group like terms together before combining)
  $7 + 2ab + 3ab = 7 + 5ab$ $-2x^2 + 7x^2 + 4x = 5x^2 + 4x$
- **Pupils can now do Exercise 10B from Pupil Book 2.2.**

Part 3

- Write a range of like and unlike terms on the board, for example:
  $5x$, $2a$, $4x2$, $3ab$, $6a$, $9y$, $2x$, $3xy$, …
- Ask pupils to identify the like terms, and to add them.
- Quiz pupils on why certain terms are unlike, for example: $5x$ and $4x^2$.

Answers

Exercise 10B

1	**a** 11h	**b** $5p$	**c** $6u$	**d** $2b$	**e** $5j$	**f** $5pr$
	g $6k$	**h** $8y^2$	**i** $5d^2$	**j** $7i$	**k** $3b$	**l** $17ab$
	m $9xy$	**n** $11p^2$	**o** $5ab$	**p** $4a^2$	**q** $6fg$	
2	**a** $8h + 5g$	**b** $2g + 8m$	**c** $8f + 10d$	**d** $11x + 5y$	**e** $6 + 2r$	**f** $4 + 2s$
	g $3c + 3$	**h** $14b + 7$	**i** $14w - 7$	**j** $6bf + 5g$	**k** $7d + 3d^2$	**l** $4t^2 + 5t$
	m $2t - 3s$	**n** $2h - 2h^2$	**o** $4y - 9w$			
3	**a** $16e + 6$	**b** $19u - 6$	**c** $4b + 3d$	**d** $7 + 7c$	**e** $6 - g$	
	f $2h + 2$	**g** $p^2 + 5$	**h** $14j^2 - 2j + 9$	**i** $4t^2 - 3t - 6$	**j** $2 - 4t + 3t^2$	
	k $q - 2p$	**l** $8 - 4e$				

4 a and d are $6x - 3y$; b and c are $5x - y$

Investigation: Four expressions

A	**a** 21, 23, 25, 27	**b** $8n + 16$	**c** 96	**d** 96
B	**a** 27, 29, 31, 33	**b** $12n$	**c** 120	**d** 120

Lesson 10.3 Expanding brackets

Learning objective

- To remove brackets from an expression

Resources and homework

- Pupil Book 2.2, pages 177–178
- Homework Book 2, section 10.3
- Online homework 10.3, questions 1–10

Links to other subjects

- **Science** – to manipulate scientific formulae

Key words

- equivalent
- expand
- multiply out

Problem solving and reasoning help

- The challenge activity at the end of Exercise 10C in the Pupil Book helps to consolidate pupils' understanding of equivalence.

Common misconceptions and remediation

- Pupils often forget to multiply all the terms in brackets. Pupils also struggle to interpret negative signs in brackets accurately. Making links to grid multiplication may be useful to help pupils make links and visualise what is happening.

Probing questions

- Multiply out these brackets and simplify the result: $(x + 5)(x - 2)$
- Explain how you multiplied out the brackets.
- Show me an expression in the form $(x + a)(x + b)$, which when expanded has:
  the x coefficient equal to the constant term, the x coefficient greater than the constant term.
- What does the sign of the constant term tell you about the original expression?

Part 1

- This is about approximations. Ask pupils what the approximate answer is to 314 × 78.
- Write the suggested answers on the board, but do not accept any calculated answers (correct answer is 24 492). Discuss the guesses, and say that guesses of more than two significant figures are not approximations.
- Show how we estimate by rounding off to a suitable number and then multiplying, for example, 300 × 80 = 24 000. Ask if we can tell if this estimate is too large or too small. (We do not know, because we rounded the 314 down and the 78 up.)
- Repeat this for the estimation of 298 × 67. This time, if we round both numbers up, our estimate is 300 × 70 = 21 000, which we know is too big, as we rounded both numbers up.
- Talk about what would happen with the estimation of 508 × 42. (We would probably round both numbers down and hence are certain of an estimate that is too small.)
- It may help pupils' understanding later in this lesson to encourage the use of the grid method for these calculations.

Part 2

- Remind pupils that the shorthand for 6 × y is 6y and the shorthand for 3 × $(t + 4)$ is $3(t + 4)$.
- On the board, write 5(3 + 4) and ask the class what value this has. If someone suggests 35, ask pupils how they calculated it. (5 × 7)
- Now show the class that we could have multiplied it out separately as 5 × 3 + 5 × 4, which is 15 + 20 = 35. So we can see that 5(3 + 4) = 5 × 3 + 5 × 4.

- This process is also called expanding.
- Show pupils how we use this technique in algebra to multiply out an expression such as $4(2m + 3)$, by showing that $4(2m + 3) = 4 \times 2m + 4 \times 3 = 8m + 12$.
- Demonstrate expanding some more brackets on the board:
  $3(2a + b) = 6a + 3b$ $\quad$ $4(x - 2y) = 4x - 8y$ $\quad$ $a(2b + c) = 2ab + ac$
  $-(a + 2) = -a - 2$ (Recall that $-(a + 2) = -1(a + 2)$)
  $-3(a - b) = -3a + 3b$ (Note that multiplying a bracket by a negative number changes the sign of each term in the brackets.)
- Point out that it will often be possible to simplify an expression further, by combining like terms, once it has been expanded. Show the class some examples:
  $2(3a + 2b) + 4a = 6a + 4b + 4a = 10a + 4b$
  $2(f + 3) + 3(2f + 7) = 2f + 6 + 6f + 21 = 8f + 27$
  $4(2d - 3e) + 3(d - 2e) = 8d - 12e + 3d - 6e = 11d - 18e$
- **Pupils can now do Exercise 10C from Pupil Book 2.2.**

Part 3

- Write up a range of expressions for pupils to expand and simplify at the board. For example:
  $3(a - 2b)$; $-4(2x - 5)$; $m(2 - n)$; $4(2i - j) - 3i$; $5(2x + 3) + 3(4x - 5)$; …

Answers

Exercise 10C

1	**a** $5p + 10$	**b** $4m - 12$	**c** $2t + 2u$	**d** $4d + 8$		
	e $5b + 25$	**f** $6j - 24$	**g** $10 + 2f$	**h** $10 - 10n$		
2	**a** $2a + 2b$	**b** $3q - 3t$	**c** $4t + 4m$	**d** $5x - 5y$		
	e $2f + 2g$	**f** $6 - 4f$	**g** $1.5h + 15$	**h** $14 - 4f$		
3	**a** $5w + 2$	**b** $3d + 10$	**c** $9h + 15$	**d** $6x + 12$	**e** $2m - 13$	**f** $4 + 3q$
4	**a** $6a + 8$	**b** $8i + 22$	**c** $5p + 4$	**d** $8d - 7$		
	e $6e + 2$	**f** $8x + 2$	**g** $8m - 3$	**h** $9u - 22$		
5	**a** $4a^2 + 6$	**b** $6x - 15$	**c** $12t + 20$	**d** $50n - 30$		
	e $30 + 12a$	**f** $8 - 12y$	**g** $60 + 80r$	**h** $35 - 10m$		
6	**a** 0	**b** 20, −24 and 4 add up to 0				

Challenge: Equivalent expressions

a, d, e, h and i are equivalent to $4x + 12$; b, c, f and g are equivalent to $4x + 8$

Lesson 10.4 Using algebraic expressions

Learning objectives

- To manipulate algebraic expressions
- To identify algebraic expressions

Resources and homework

- Pupil Book 2.2, pages 179–182
- Homework Book 2, section 10.4
- Online homework 10.4, questions 1–10

Links to other subjects

- **Science** – to manipulate scientific expression.

Key word

- manipulate

Problem solving and reasoning help

- This lesson makes links to perimeter and area to help pupils gain a visual image of what they are doing. It also uses arithmogons, which help pupils to see the power of algebra in terms of generalisation, so that pupils will start to manipulate expressions using inverse operations to them find the solution to more complex problems.

Common misconceptions and remediation

- Difficulties with substitution into formulae often come from a failure to grasp the basics of BIDMAS and negative numbers, along with an inability to recognise that letters represent variables. When this is the case, pupils tend to want to substitute specific values. Provide lots of opportunity for pupils to see this in action in familiar contexts, before moving on to more complex or abstract examples.

Probing questions

- What are the important steps when substituting values into this expression or formula?
- What would you do first? Explain why.
- How would you continue to find the answer?
- When you substitute $x = 3$ and $y = 4$ into the formula $w = xy + 3xy$ you get 48. Can you make up other formulae that will give an answer of 48 when $x = 3$ and $y = 4$?

Part 1

- Tell pupils that you will work with complements of 10. Complements of numbers are the numbers needed to add to numbers to make 10. For example, the complement of 7 is 3.
- Start with 2.5 (complement of 7.5), and ask for the complements of decimal numbers such as: 3.5, 8.5, 4.5, 7.2, 8.3, 5.4, 1.7, 9.4. After a few, stop and ask pupils how they are trying to find the answer. Some will add on the decimal part to reach the next integer and then build up to 10; others will use a subtraction method. Explore the diversity of methods, and encourage pupils to use the technique that is clear to them. Ask pupils for some examples.
- Then move onto fractions, three-and-a-half, four and two-eighths. Again, stop after a few and ask pupils how they are doing this – is it the same method as they used for the decimals?
- Finish by asking again for complements of 10, but mix decimals with fractions, for example: 2.4, three and five-eighths, 6.3, five and two-fifths, 8.1, seven and five-sixths ….

Part 2

- On the board, draw the shape shown and ask pupils what its perimeter is. Add the lengths to give $a + 3b + 2a + 2b + 1$, which simplifies to $3a + 5b + 1$.

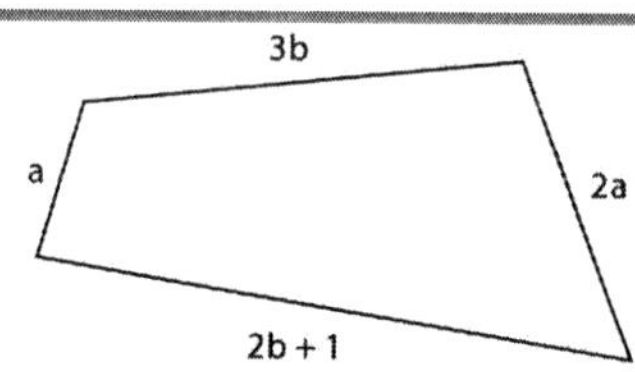

- The class can now do questions 1 and 2 of Exercise 10D in Pupil Book 2.2.
- Draw the next shape on the board. Ask pupils what its area is.

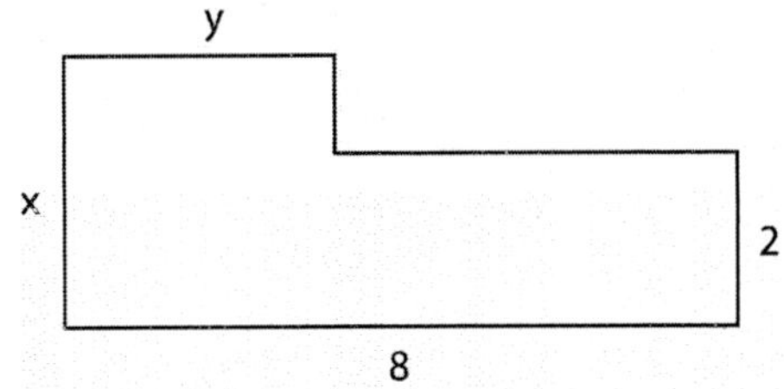

- Show how the shape can be split into two rectangles, parts A ($y \times x$) and B [$2 \times (8 - y)$].
- Ask pupils for the area of part A. (xy)
- Ask pupils for the area of part B. [$2(8 - y) = 16 - 2y$]
- Show that the total area of the shape is given by: $xy + 16 - 2y$.
- The class can now do the rest of Exercise 5D from Pupil Book 2.2.

Part 3

- Talk to the class about how the work they are doing with shapes here looks quite different from the work that involved finding the actual length of perimeters and the actual areas of shapes. Explain that there are times when we have to use algebra like this to help us find the actual perimeter or area.
- For example, if we knew that the shape shown had an area of 100 m2, how could we find x? The answer to this will not be dealt with until Year 10 or Year 11.

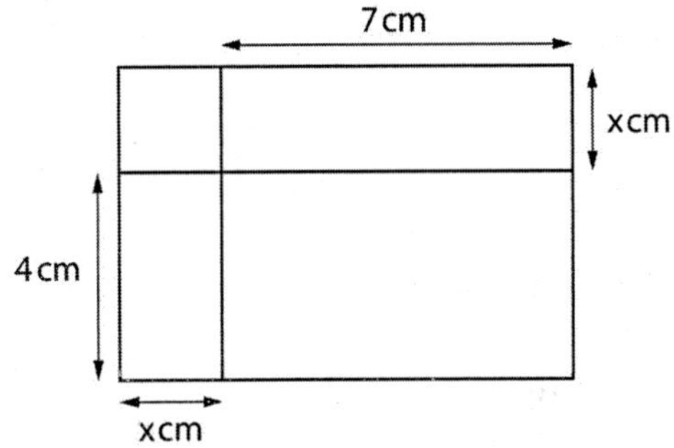

Answers

Exercise 10D

1 **a** $4t + 2$ **b** $6w$ **c** $6m + 2$

2 **a** $t(t + 1)$ **b** $2w^2$ **c** $m(2m + 1)$

3 **a** $3t + 6$ **b** $4k + 8$ **c** $4c + 16$ **d** $3a + 24$ **e** $5x + 10$

4 Expand the brackets in each case

5 **a** $9x$ **b** $10ab$ **c** $16t^2$

6 **a i** $4x$ **ii** $4(x + 4)$ **iii** $2(x + 1)$ **iv** $2(x + 3)$
b Expand brackets and simplify
c i $2x + 8$ **ii** $2x + 16$ **iii** $2x + 6$ **iv** $2x + 10$
d Expand brackets and simplify

7 **a i** $5(a + 2)$ **ii** $4(a + 2)$ **iii** $9a$ **iv** $9(2a + 2)$ or an equivalent expression
b i $2a + 14$ **ii** $2a + 12$ **iii** $2a + 18$ **iv** $4a + 22$

8 **a** $2x + 21$ **b** $2(x - 3) + 27$ **c** Expand brackets and simplify

9 Own demonstration showing $3x + 8 + 3x + 4 = 6x + 12$

10 Own demonstration showing $4x + 6 + 4x + 10 = 4x + 16$

11 Own demonstration showing $2x + 9 + 3x + 6 = 5x + 15$

12 Own demonstration showing $3x - 5 + 3x - 1 = 6x - 6$

Challenge: Fill in the bricks

There are many possible answers.

Lesson 10.5 Using index notation

Learning objective

- To write algebraic expressions involving powers

Resources and homework

- Pupil Book 2.2, pages 182–183
- Homework Book 2, section 10.5
- Online homework 10.5, questions 1–10

Links to other subjects

- **Science** – to be able to work with very large and very small numbers

Key word

- index
- power

Problem solving and reasoning help

- The challenge at the end of Exercise 10E in the Pupil Book looks at algebraic multiplication tables using powers. You could make links here to work on symmetry groups.

Common misconceptions and remediation

- Pupils struggle to identify what constitutes 'like' when powers are included. Provide plenty of opportunity for pupils to compare and discuss what are and what are not like terms.

Probing questions

- Explain why you cannot simplify this expression: $3x^2 + x + 2$.

Part 1

- Ask the class if they can divide 420 by 15 in their heads without writing down anything or using a calculator. It is most unlikely that anyone will be able to answer this (28). However, if someone does give the answer, ask that person to tell the class how she or he did it.
- Explain that they can work out the answer mentally in various ways: 1. Divide by 3 and then divide by 5 (420 ÷ 3 = 140, then ÷ 5 to give 28). 2. Divide by 5 and then by 3 (420 ÷ 5 = 84, then ÷ 3 to give 28). 3. Divide by 30, and then double (420 ÷ 30 = 14, then double to give 28). Explain how this works, that to divide by 10 and then double is how you can divide by 5.
- Ask pupils to try out the various techniques by writing numbers on the board and asking them to divide each number by 15. Use the following numbers: 345 (23), 465 (31), 240 (16), 390 (26), 645 (43), 525 (35), 195 (13), 360 (24), 495 (33), 630 (42), 270 (18), 570 (38).
- After most pupils have used one of the techniques correctly and understand it, ask how they can tell if a number will divide by 15 before we start dividing. It must be both a multiple of 5 (end in a 5 or a 0) and a multiple of three (digits add up to a multiple of 3).

Part 2

- Ask for a quick way of writing 2 × 2 × 2 (2^3) and recall index notation (base = 2, index = 3).
- Replace 2 by a and ask how to write $a \times a \times a$ concisely (a^3).
- Provide more examples for pupils to abbreviate: $m \times m \times m \times m \times m = m^5$; $2a \times a = 2a^2$; $4m \times 3m \times 2m = 24m^3$; $f + f + f + f + f \times f \times f \times f = 4f + f^4$
- Ask the class to tell you the difference between c^5 and $5c$.
- **Pupils can now complete Exercise 10E from Pupil Book 2.2.**

Part 3

- On the board, write $3(4t - 2)$ and ask pupils to tell you how to expand it. From previous lessons they should remember that this is: $12t - 6$.
- Now write on the board $m(4m - 2)$ and ask pupils if they can expand this product.

- From pupils' suggestions, it is likely that that you will get the correct answer of $4m^2 - 2m$; you need to show why the wrong suggestions are not correct. Ensure that each pupil sees the process as identical to the one above, in which numbers are used instead of variables.
- On the board, go through a few more similar expansions with the class.
- **More able** pupils could try more complex examples such as: $m(m + 3p) - p(2m - 5p)$, and go through the expansion carefully to get $m^2 + 3mp - 2mp + 5p^2$, especially the $-p \times -5p = +5p^2$ step, and then simplify to: $m^2 + mp + 5p^2$.

Answers

Exercise 10E

1 **a** a^4 **b** r^3 **c** b^5 **d** m^6 **e** $12a^2$ **f** $2p^2$ **g** $12g^3$ **h** $24k^4$
2 **a** $5f$ **b** w^4 **c** $7c$ **d** k^4 **e** $6d$
3 $5j = j + j + j + j + j$ and $j^5 = j \times j \times j \times j \times j$
4 6, 12, 15; 8, 64, 125; 24, 192, 375
5 **a** a^2b **b** $6xy^2$ **c** t^2u^2 **d** $2d^2c$ **e** $2ab^2$ **f** $2xw^2$ **g** $4t^2u$ **h** $6e^2f$
6 **a** 75 **b** 250
7 **a** $4x^2$ **b** y^3 **c** $2t^3$ **d** $4k^3$ **e** $6n^3$
8 **a** t^4 **b** $6t^4$ **c** t^4 **d** $30t^4$

Challenge: Matching multiples

First row: $2a$, $2b$, $2a^2$, $2b^2$, $2ab$; second row: a^2, ab, a^3, ab^2, a^2b; third row: ab, b^2, a^2b, b^3, ab^2

Review questions

(Pupil Book pages 184–185)

- The review questions will help to determine pupils' abilities with regard to the material within Chapter 10.
- These questions also draw on the maths covered in earlier chapters of the book to encourage pupils to make links between different topics.
- The answers are on the next page.

Mathematical reasoning – Writing in algebra

(Pupil Book pages 186–187)

- Pupils often struggle to decode everyday language into mathematics. This activity has been designed to develop pupils' confidence and fluency with algebraic notation, and gives them the opportunity to practise this in a range of contexts.
- In preparation for this activity discuss the fact that mathematicians have not always been able to use algebra to do mathematics. Algebra as we know it really began to develop around 1500 but did not fully replace words until the 17th century.
- Say that in this activity, pupils will have the opportunity to experience the benefits of being able to decode some real-life problems that are expressed in words into algebra.
- You could prepare the class for this by doing a 'think of a number' activity. For example, write a mathematical expression in words on the board such as:
- 'Take a number and multiply by 12, then subtract 7 from this number. This equals 53.'
- Ask pupils to write this using algebra, and check their solutions. Highlight any conventions about notation. Then, reflecting on what you have just told pupils about the history of algebra let them discuss the advantages of writing this expression algebraically.
- Depending on the class, you might want to change the expression to support or challenge pupils. If necessary, provide both the written and the algebraic version for pupils to compare and discuss. Go through the perimeter example in the 'Writing in algebra' activity in the Pupil Book. Then pupils can work through questions 1 to 9.
- Encourage **more able** pupils to research further about the development of algebra. To reinforce the power of algebra, you could also ask them to write some of the axioms of number algebraically. For example, the commutative law of multiplication:
  'Take a number and multiply it by a second number. This is the same as taking the second number and multiplying it by the first number.'
- You could ask pupils to write their own word problems, which they can then write algebraically. Encourage pupils to include real-life contexts including financial examples.

Answers to Review questions

1 **a** $2 + 3t$ **b** $20d$ **c** $2a + 4b$ **d** $\frac{1}{2}c^2$
e $2r - 12$ **f** $20 - 3q$ **g** $\frac{2a}{b}$ **h** $6km$

2 **a** $4h$ **b** $a + 3b$ **c** $4p - 2$ **d** $7x - 8$
e $4x^2$ **f** $7 - 2a$ **g** $2d + 3d^2$ **h** $r - 8 + 5r^2$

3 **a** $3t + 3$ **b** $7x$ **c** $6m - 2$

4 **a** $2x + 20$ **b** $4y + 2$ **c** $4m - 6$

5 **a** $10x$ **b** $y(y + 1)$ **c** $m(m - 3)$

6 **a** $10ab$ **b** $b(a + 10)$ or $ab + 10b$

7 **a** $2a + 16$ **b** $3f - 18$ **c** $\frac{1}{4}t + 5$

8 **a** $\frac{w + x + y + z}{4}$ **b** 20

9 $6t^3$

10 **a** $4x^2$ **b** $6x + 32$ **c** $2x^2 + 16x$ or an equivalent expression

Answers to Mathematical reasoning – Writing in algebra

1 Power = vc
2 Distance = st
3 Volume = $\frac{1}{3}sh$
4 Area = $1.72l^2$
5 Volume = $0.79d^2h$
6 Height = $5t^2$
7 bmi = $\frac{m}{h^2}$
8 Kinetic energy = $\frac{1}{2}s^2m$
9 Cost = $90d + nc$

11 Congruence and scaling

Learning objectives

- How to recognise congruent shapes
- How to enlarge a shape by a scale factor
- How to use shape and ratio
- How to use scales in drawings and maps

Prior knowledge

- How to reflect a 2D shape in a mirror line
- How to rotate a 2D shape about a point
- How to translate a 2D shape
- How to use ratio

Remind pupils about the different transformations they met in Year 7.

Context

- This chapter first shows pupils how to recognise congruent shapes, and then reminds them how to transform a 2D shape by a reflection, a rotation and a translation. Pupils are then introduced to combining transformations and enlarging a 2D shape about a given centre of enlargement and scale factor. Finally, pupils are shown how to use scales and ratio in drawings and maps.

Discussion points

- What do you look for when deciding if two triangles are congruent?
- Can you use congruence to show that the diagonals of a rhombus cut each other at right angles?
- How could you find the scale factor and centre of enlargement from a completed enlargement?

Associated Collins ICT resources

- Chapter 11 interactive activities on Collins Connect online platform
- *Using enlargement* and *Using proportion and scale factors* videos on Collins Connect online platform
- *Construction views* and *Large bike* Wonder of Maths on Collins Connect online platform

Curriculum references

Develop fluency

- Use language and properties precisely to analyse numbers, algebraic expressions, 2D and 3D shapes, probability and statistics

Reason mathematically

- Extend and formalise their knowledge of ratio and proportion in working with measures and geometry, and in formulating proportional relations algebraically

Ratio, proportion and rates of change

- Use scale factors, scale diagrams and maps
- Use ratio notation, including reduction to simplest form
- Understand that a multiplicative relationship between two quantities can be expressed as a ratio or a fraction

Geometry and measures

- Use the standard conventions for labelling the sides and angles of triangle ABC, and know and use the criteria for congruence of triangles
- Identify properties of, and describe the results of, translations, rotations and reflections applied to given figures
- Identify and construct congruent triangles, and construct similar shapes by enlargement, with and without coordinate grids

Fast-track for classes following a 2-year scheme of work

- Pupils will have met some of the basic concepts in this chapter. If they can demonstrate that they are confident and fluent with these basic concepts, pupils can move on to the more challenging questions at the end of each exercise.

Lesson 11.1 Congruent shapes

Learning objectives

- To recognise congruent shapes

Links to other subjects

- **Art** – to use congruence within artworks
- **Design and technology** – to use congruent shapes in design such as bridges or textile patterns

Resources and homework

- Pupil Book 2.2, pages 189–192
- Homework Book 2, section 11.1
- Online homework 11.1, questions 1–10

Key words

- congruent
- congruent triangles
- transformation

Problem solving and reasoning help

- **MR** questions 5 and 6 in Exercise 11A of the Pupil Book require pupils to create congruent shapes, either by joining two congruent shapes or by splitting a shape into smaller congruent shapes. A basic understanding of transformations will help pupils to answer these questions. If pupils struggle with **PS** question 7, briefly go through rotational symmetry.

Common misconceptions and remediation

- Pupils often do not realise that you can test for congruence by placing one shape on top of the other. Encourage the use of tracing paper to do this. Also reinforce the fact that shapes can have different orientations and still be congruent.

Probing questions

- What will stay the same and what will change a shape when you:
  - reflect
  - translate
  - rotate.
- How can you tell if two images are congruent?
- Can you list all the ways in which two triangles can be congruent?
- If two triangles have identical angles, are the shapes congruent?

Part 1

- Imagine a square. Now imagine another square of exactly the same size, which touches the first one along one of its sides. What shape are you thinking of? (A rectangle)
- Imagine an equilateral triangle. Imagine another equilateral triangle of exactly the same size, which touches the first one along one side. What shape are you thinking of? (A rhombus)

Part 2

- Remind pupils about the different transformations they met in Year 7. Draw the following diagrams on the board or on an OHT:

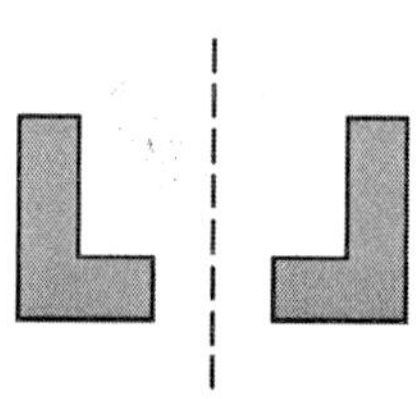

A reflection in a mirror line

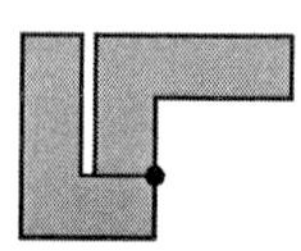

A rotation about a point

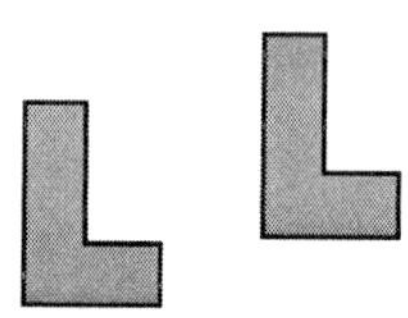

A translation

- Ask pupils to describe what happens to the shapes after any of the three transformations. They should remember that the object and the image are the same shape and size.
- Write: 'Two shapes are congruent if they are exactly the same shape and size. Reflections, rotations and translations all produce images that are congruent to the original object.'
- Have prepared sets of various congruent triangles and quadrilaterals made from card. With the class working in groups, let them sort the shapes into congruent pairs.
- Make sure pupils understand that for each pair of congruent shapes the corresponding sides and angles are equal.
- **Pupils can now do Exercise 11A from Pupil Book 2.2.**
- Card, scissors and square-dotted paper may be required for this exercise.

Part 3

- Ask pupils to explain what congruent shapes are. Invite pupils to draw two congruent shapes on the board or on individual whiteboards to show to the rest of the class.
- Ask pupils to describe what would happen if two shapes are the same, but are different in size. This should lead to a discussion on enlargements.

Answers

Exercise 11A

1 **a** yes; **b** no; **c** yes; **d** yes; **e** no; **f** yes

2 **a** and **e**; **b** and **j**; **c** and **k**; **d** and **f**

3 **a** and **c**

4 **a** Two different isosceles triangles, two different parallelograms, a rectangle and a kite

b A parallelogram and a rhombus

c A rhombus

5

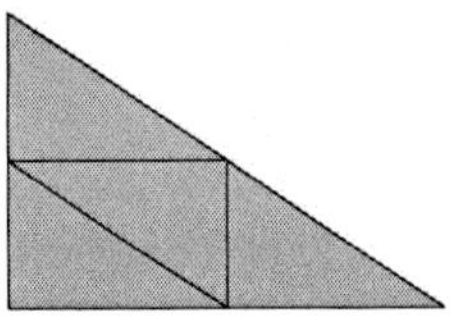

6

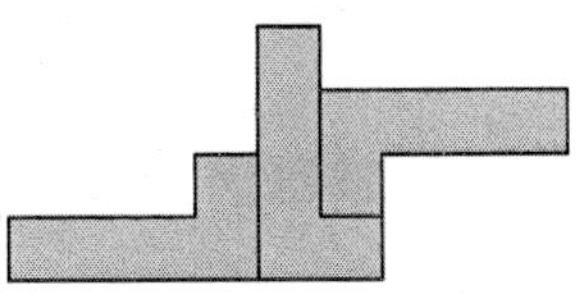

7 **a** Examples of two congruent shapes:

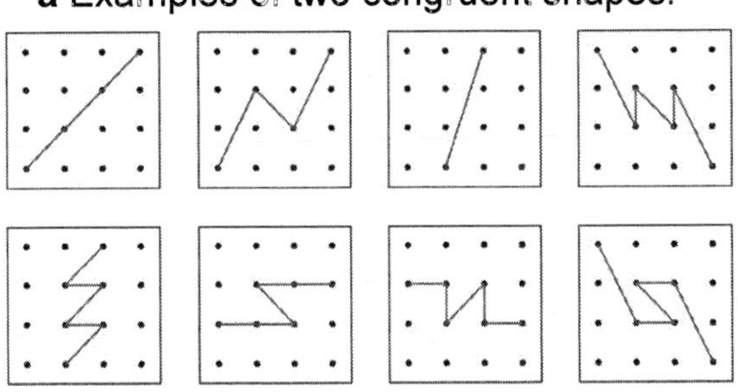

b Examples of four congruent shapes:

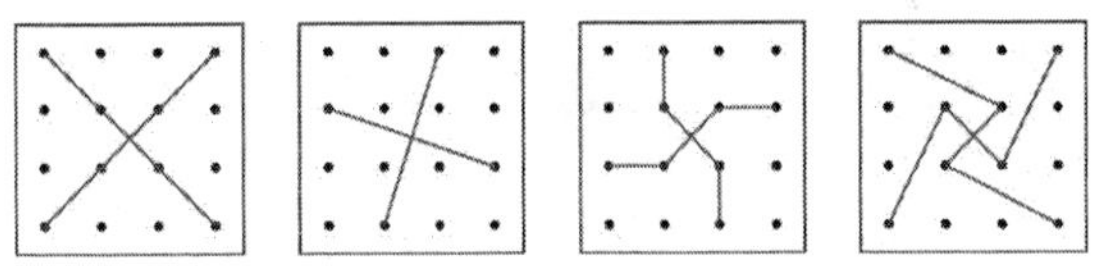

8 **a** $\angle C = \angle D$, BC = DE, AC = DF (SAS)

b GH = KL, GI = JL, HI = JK (SSS)

c $\angle N = \angle R$, $\angle O = \angle P$, NO = PR (ASA)

d $\angle B = \angle X$, $\angle C = \angle Y$, BC = XY (ASA)

Reasoning: Combined transformations

The following are possible examples for a combined transformation:

1 **a** A reflection in the *y*-axis followed by a translation of 6 units down

b A rotation of 90° anticlockwise about the origin followed by a translation of 1 unit down

c A reflection in the *x*-axis followed by a translation of 7 units left and 5 units down

d A rotation of 90° clockwise about the origin followed by a translation of 1 unit left and 6 units up

e A rotation of 90° anticlockwise about the origin followed by a translation of 6 units left and 6 units up

Lesson 11.2 Enlargements

Learning objectives

- To enlarge a 2D shape by a scale factor

Resources and homework

- Pupil Book 2.2, pages 193–197
- Homework Book 2, section 11.2
- Online homework 11.2, questions 1–10

Links to other subjects

- **Art** – to enlarge a drawing using a grid
- **Design and technology** – to enlarge a design by a scale factor

Key words

- centre of enlargement
- ray
- enlargement
- scale factor

Problem solving and reasoning help

- **MR** question 4 in Exercise 11B if the Pupil Book requires pupils to enlarge a shape with the centre of enlargement inside the shape. Pupils often find this type of question difficult, as they are not sure where to draw the lines. Remind pupils to follow the method they have been taught. **MR** question 6 requires pupils to draw axes and construct the shape to be enlarged from the given coordinate points. Pupils sometimes struggle to do this accurately. In pairs, ask pupils to check each other's axes and then shapes to make sure they have been drawn correctly. Help **less able** pupils with drawing correct axes.

Common misconceptions and remediation

- Pupils often use an incorrect point as the centre of enlargement, or they just enlarge the shape without reference to the given point.
- Sometimes pupils do not enlarge all the lines, or they enlarge by an incorrect scale factor. Remind pupils that the shape produced after an enlargement will look similar to the original; it will just be a bigger or smaller version.

Probing questions

- What changes when you enlarge a shape?
- What stays the same when you enlarge a shape?
- What information do you need to be given in order to enlarge a shape?

Part 1

- On the board, write or place a multiplication grid similar to the one shown. Ask individual pupils to complete any cell in the multiplication grid. Then ask if they have any particular strategies for working out their answers.
- Discuss some strategies. For example: leave out the decimal point, then multiply and the answer will have one decimal place; to multiply by 4, double the number and double again.

	× 2	**× 3**	**× 4**
1.2			
2.5			
3.4			
5.6			
7.9			

Part 2

- Ideally, this work should cover two lessons. Try to have enlargement examples ready to show the class. Enlargements of photographs will also help pupils to understand the work.
- Remind pupils that the three transformations they have met so far (reflections, rotations and translations) do not change the size of an object. They will now look at a transformation that does change the size of an object – an *enlargement*.

- Draw the diagram as shown on the board.

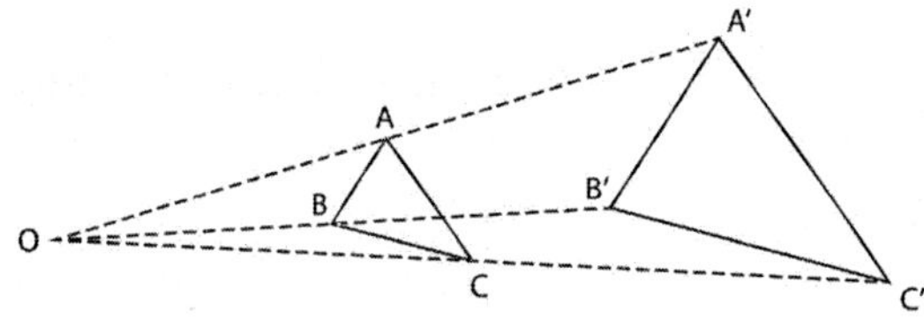

- Explain that all the sides of ΔA′B′C′ are twice as long as the sides of ΔABC and that OA′ = 2 × OA, OB′ = 2 × OB and OC′ = 2 × OC. ΔABC has been enlarged by a scale factor of two about the centre of enlargement O to give the image A′B′C′. The dotted lines are called guidelines or rays for the enlargement. Explain that to enlarge a shape a centre of enlargement and a scale factor are needed.

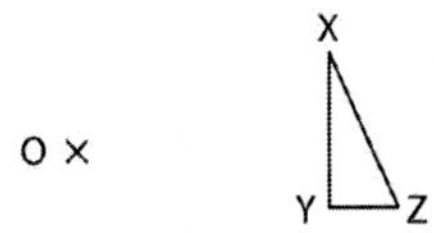

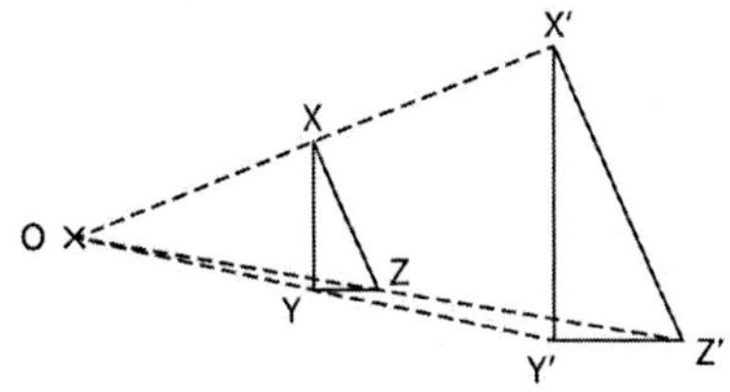

- Show pupils how to enlarge triangle XYZ by a scale factor of two about the centre of enlargement O (shown right). Draw rays OX, OY and OZ. Measure the length of the three rays and multiply each length by two. Extend each ray to these new lengths measured from O and plot the points X′, Y′ and Z′. Join X′, Y′ and Z′. ΔX′Y′Z′ is the enlargement of ΔXYZ by a scale factor of two about the centre of enlargement O.

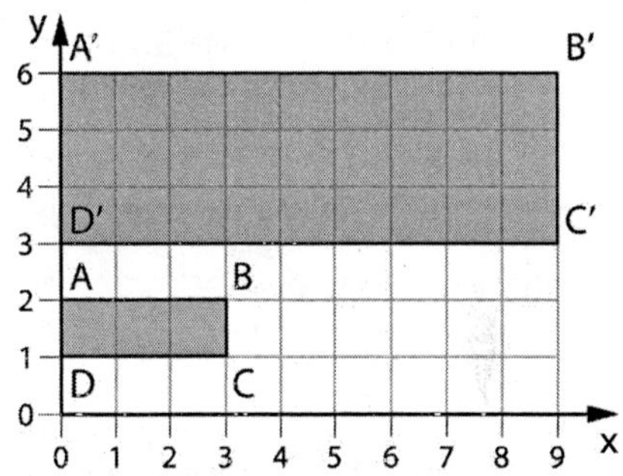

- Show pupils how to enlarge a shape about the origin on a coordinate grid. The rectangle ABCD on the coordinate grid shown is enlarged by a scale factor of three about the origin O to give the image rectangle A′B′C′D′:
- The coordinates of the object are: A(0, 2), B(3, 2), C(3, 1) and D(0, 1).
- The coordinates of the image are: A′(0, 6), B′(9, 6), C′(9, 3) and D′(0, 3).
- Note that if a shape is enlarged by a scale factor about the origin on a coordinate grid, the coordinates of the enlarged shape are multiplied by the scale factor.
- **Pupils can now start Exercise 11B from Pupil Book 2.2.**

Part 3

- Ask which properties of a shape change after an enlargement and which properties remain the same. For example: length of sides change; area changes; all angles stay the same size.

Answers

Exercise 11B

1 Check pupils' answers

2 **a** Vertices at (8, 6), (8, 2), (4, 2) **b** Vertices at (4, 6), (8, 4), (4, 2), (0, 4)
c Vertices at (3, 9), (6, 9), (6, 6), (9, 6), (9, 9), (12, 9), (12, 3), (3, 3)
d Vertices at (0, 8), (8, 8), (8, 12), (12, 6), (8, 0), (8, 4), (0, 4)

3 Vertices at (8, 12), (10, 8), (8, 2), (6, 8)

4 **a** A′(3, 7), B′(7, 7), C′(7, 3), D′(3, 3) **b** A″(2, 8), B″(8, 8), C″(8, 2), D″(2, 2)
c A‴(1, 9), B‴(9, 9), C‴(9, 1), D‴(1, 1)
d For example, the *x*-coordinate and the *y*-coordinate are the same or they add up to 10

5 **a** 2 **b** (9, 1)

6 **b** 4 cm^2 **c** 16 cm^2 **d** 36 cm^2 **e** 64 cm^2
f The area scale factor is the square of the scale factor **g** yes

Activity: Enlarged stickman

Check pupils' posters

Lesson 11.3 Shape and ratio

Learning objective

- To use ratio to compare lengths, areas and volumes of 2D and 3D shapes

Resources and homework

- Pupil Book 1.2, pages 197–200
- Intervention Workbook 3, pages 40–42
- Homework Book 2, section 11.3
- Online homework 11.3, questions 1–10

Links to other subjects

- **Science** – to work with volumes and ratio in experiments
- **Art** – to enlarge a 3D shape

Key word

- hectare

Problem solving and reasoning help

- Make sure pupils have an understanding of how to calculate the surface area and volume of a cuboid before they tackle this question. Model some examples for **less able** pupils.

Common misconceptions and remediation

- Pupils will need reminding of how to work with ratio. Therefore, plan a revision exercise to remind pupils how to simplify ratios.

Probing questions

- The ratio of two squares is 1 : 9. The larger square has an area of 36 cm^2. What is the area of the smaller square?
- Two cubes have a ratio of 1 : 3. The volume of the smaller cube is 27 cm^3. How would you calculate the length of the side on the larger cube?

Part 1

- Write the following ratios on the board:

1 3 : 6	**2** 5 : 20	**3** 8 : 12	**4** 15 : 25	**5** 24 : 36
6 18 : 24	**7** 14 : 35	**8** 30 : 50	**9** 40 : 100	**10** 45 : 60

- Ask pupils to give the ratios in their simplest form. Pupils can either give oral answers or show them on individual whiteboards. Answers:

1 1 : 2	**2** 1 : 4	**3** 2 : 3	**4** 3 : 5	**5** 2 : 3
6 3 : 4	**7** 2 : 5	**8** 3 : 5	**9** 2 : 5	**10** 3 : 4

Part 2

- The lesson is to show pupils how *ratio* can be used to solve problems that involve *length*, *area* and *volume*. The idea is for pupils to grasp that the ratio of lengths, areas and volumes is different when comparing 2D and 3D shapes.
- Explain the following three examples to the class.
- Remind pupils that a ratio does not have units, and is always given in its *simplest form.*

1. Find the ratio of the length of line segment AB to the length of line segment XY:

A ———————— B   X ———————————— Y

80cm   1.2m

Since the lengths are in mixed units, change them to the smallest unit before simplifying the ratio. The ratio is 80 cm : 1.2 m = 80 cm : 120 cm = 2 : 3.

2. Find the ratio of the area of rectangle A to the area of rectangle B:

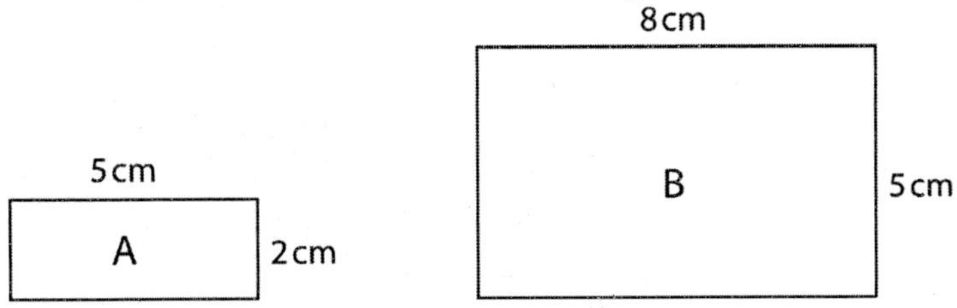

The ratio is 10 cm^2 : 40 cm^2 = 1 : 4.

3. Find the ratio of the volume of cube A to the volume of cube B:

edge length 2 cm   edge length 3 cm

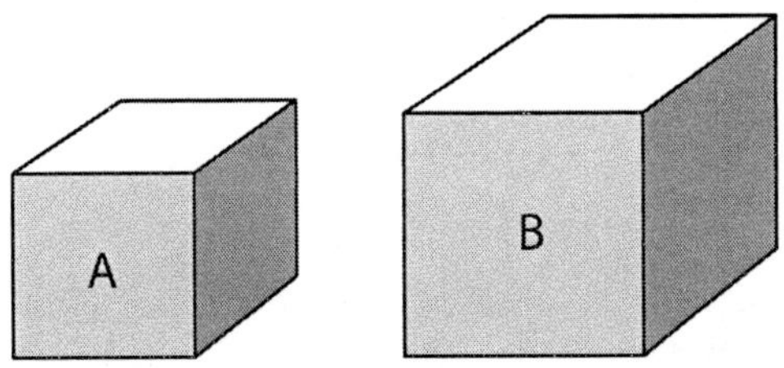

The ratio is 8 cm^3 : 27 cm^3 = 8 : 27.

- **Pupils can now do Exercise 11C from Pupil Book 2.2.**

Part 3

- Ask pupils to write a brief summary of what they have learnt during the lesson.
- Discuss the connections between ratio, fraction, percentage and enlargement scale factor.

Answers

Exercise 11C

1	**a** 2 : 5	**b** 1 : 10	**c** 4 : 5	**d** 1 : 5	**e** 1 : 4
2	**a** 1 : 3	**b** 1 : 3	**c** 1 : 9		
3	**a i** 1 : 2 : 3	**ii** 1 : 2 : 3	**iii** 1 : 4 : 9	**b** They are enlargements of each other	
4	**a** 1 : 1	**b** 1 : 5	**c** 2 : 5	**d** 1 : 2	
5	**a** 1 : 8	**b** $\frac{1}{8}$			
6	**a** 1200 m^2	**b i** 30 000 m^2		**ii** 3 hectares	
	c 1 : 5	**d** 1 : 25	**e** $\frac{1}{25}$		
7	**a** 24 litres	**b** 18 litres	**c** 3 : 4		
8	**a** 1 : 2	**b** 1 : 4	**c** 1 : 8	**d i** 1 : 9	**ii** 1 : 27

Activity: Paper sizes

A A5: 210 mm × 148 mm, A4: 297 mm × 210 mm, A3: 420 mm × 297 mm

B They are the same

C 1 : 1.4 (actual ratio 1 : $\sqrt{2}$), they are the same ratio

Lesson 11.4 Scales

Learning objective

- To understand and use scale drawings
- To know how to use map ratios

Resources and homework

- Pupil Book 2.2, pages 200–203
- Intervention Workbook 2, pages 43–45
- Intervention Workbook 3, pages 40–42
- Homework Book 2, section 11.4
- Online homework 11.4, questions 1–10

Links to other subjects

- **Geography** – to work with maps and navigation
- **Design and technology** – to use scales to make scale drawings

Key words

- map ratio
- scale
- scale drawing

Problem solving and reasoning help

- The **MR** questions in Exercise 11D of the Pupil Book require pupils to understand map scales and ratio. Make sure pupils understand how to design a scale in order to make a scale drawing that will fit on the page. Pupils will also need to be familiar with working out actual distances from a map scale in the form 1 : *x*. Model some examples if necessary.

Common misconceptions and remediation

- Pupils will occasionally mix units when working with map scales. Make sure pupils understand that units must be the same before writing a ratio.

Probing questions

- How would you decide on the scale you are going to use when making a scale drawing of a large object such as a football field?
- Explain how you would work out how to calculate the actual distance between two objects on a map with a scale of 1 : 10 000.

Part 1

- Pupils will practise estimating length. Pupils will need rulers and a tape measure. This activity is best done using individual whiteboards, but pupils can give their answers orally.
- Ask a pupil to select an object in the classroom. Ask the rest of the class to estimate the length of the object and to write their answer on their whiteboards.
- Ask the first pupil to measure the actual length of the object.
- Pupils can now show their estimates. Some discussion of the units used may be a useful exercise. Pupil whose estimate is closest to the actual length wins a point.
- Repeat the activity with a different pupil selecting another object.

Part 2

- Explain to pupils that this lesson is about using and making *scale drawings*.
- On the board, make a scale drawing to show the *plan* dimensions of, for example, a classroom, the school hall or gym. Explain the importance of choosing a sensible *scale*.
- It might be worth pointing out that if you double the

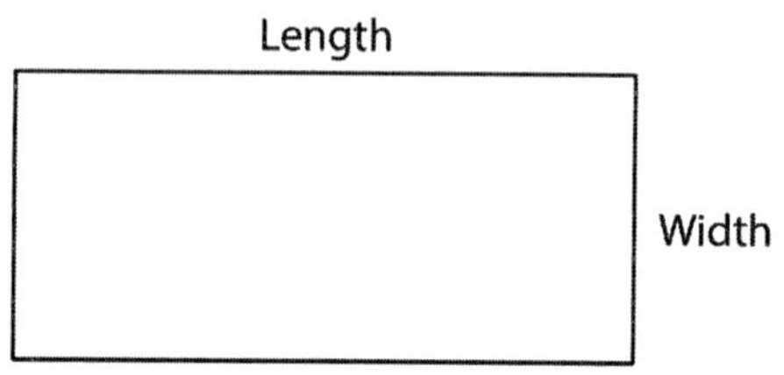

Scale 1 cm to 2 m

scale, the scale drawing will be halved in size. In the example, doubling the scale gives 1 cm to 4 m. Show pupils how to find the actual length and width of the room by using the scale.

- *Map scales*: Explain that *maps* are *scale drawings* used to represent areas of land in subjects such as surveying and geography. Show pupils various maps and the different *map scales* that are used. Show pupils a map with a map scale of, for example, 1 cm to 10 km.
- Show pupils how to work out the actual direct distance between two places on the map. Direct distance is 'as the crow flies'. For example, if the direct distance between A and B on a map is 4 cm, and the scale is 1 cm to 10 km, then the actual direct distance between A and B is 4 × 10 km = 40 km.
- Explain to pupils that an alternative to a map scale is a *map ratio.* A map ratio is the ratio of a distance on the map to the actual distance it represents on the ground. Map ratios are given in the form 1 : *x* and have no units. A map ratio is also known as a *scale factor.*
- For example, the map scale might be 1 cm to 5 km. Now 1 km is 100 × 1000 cm. So the scale is 1 cm to 500 000 cm. Then the map ratio can be given as 1 : 500 000.
- **Pupils can now do Exercise 11D from Pupil Book 2.2.**

Part 3

- Ask pupils to explain the two ways of showing distance on a map.
- Ask individuals to change various map scales to map ratios, for example: 1 cm to 100 m, 1 cm to 2 km, 1 cm to 50 km. (Respectively: 1 : 10 000, 1 : 200 000, 1 : 5 000 000)

Answers

Exercise 11D

1	**a** 20 m	**b** 50 m	**c** 35 m	**d** 78 m	**e** 63m	
2	**a** 25 m	**b** 15 m	**c** ≈ 29 m			
3	**a** 1 cm to 2 m	**b** 6 m	**c** 48 m^2			
4	**a** 16 cm	**b** 6 cm	**c** 2 cm	**d** 3 m	**e** 2.5 m	**f** 1.2 m
5	**a i** 6 m by 4 m	**ii** 4 m by 2 m	**iii** 6 m by 4 m	**iv** 5 m by 4 m	**b** 88 m^2	
6	For example 1 cm to 10 yards					
7	**a** 1 km	**b** 1.4 km	**c** 2.25 km	**d** 1.75 km		

Reasoning: Map ratios

A 1 : 10 000
B 1 : 12 500
C 1 : 100 000
D 1 : 50 000
E 1 : 1 000 000

Review questions

(Pupil Book pages 204–205)

- The review questions will help to determine pupils' abilities with regard to the material within Chapter 11.
- These questions also draw on the maths covered in earlier chapters of the book to encourage pupils to make links between different topics.
- Answers are on the next page.

Problem solving – Photographs

(Pupil Book pages 206–207)

- This activity consolidates topics previously covered on extracting data, area and ratio.
- Draw the squares and rectangles on the board or OHP as shown.
- Ask pupils to answer these questions:
  - o Which two shapes have the same perimeter? (E and G)
  - o Which two shapes have the same area? (D and F)
  - o Which two shapes have the same value for their perimeter and area? (B and C)

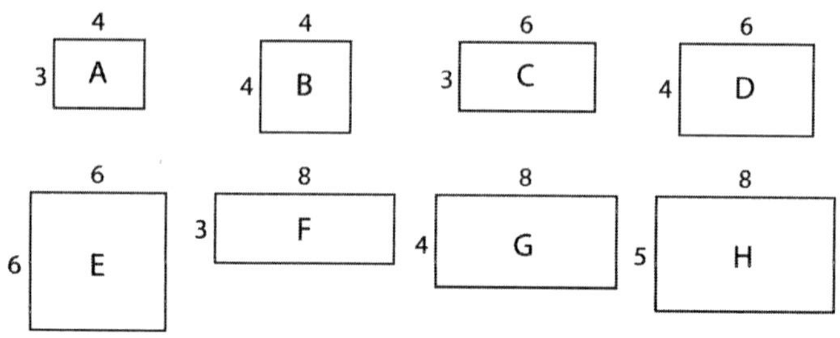

- Check pupils' answers after each question to ensure that they understand the difference between perimeter and area.
- Encourage pupils to suggest questions that might arise from the data in the activity, for example, the use of inches for the dimensions of the prints. Discuss the imperial units that are still being used. Examples are weighing vegetables at market stalls (pounds and ounces) and human weight (stone).
- Discuss the reasons why shops encourage us to buy in bulk.
- It is important for pupils to realise that when we refer to 'photograph enlargements', they may not actually be true mathematical enlargements. It may be useful to revise enlargements and scale factors before pupils start question 4 of the activity.
- Discuss 'best buys' with pupils. Do people really compare prices in different shops? Ask pupils for their views. This may well lead to some discussion about the influence of advertisements on TV and on the internet.
- Before starting question 6 on golden rectangles, it may be useful to revise simplifying ratios.
- Golden rectangles appear in paintings and buildings, both ancient and modern. For example, the painting by Salvador Dali, 'The Sacrament of the Last Supper' is a golden rectangle and the exterior of the Parthenon on the Acropolis in Athens is based on the golden rectangle.
- Direct pupils to the internet where there is a wealth of information on golden rectangles and the golden ratio, Ø.

  $Ø = \frac{\sqrt{5}+1}{2} = 1.618\ 039\ 887\ 498\ 948\ 482\ 045\ 868\ 343\ 656\ 381\ 177\ 203\ 091\ 798\ 057\ 6 \ldots$
- Let pupils work through questions 1 to 6 in the activity
- You could end by asking the class for the different areas of mathematics that they have used during the lesson. Individual pupils could write these on the board or on a flip-chart.

Answers to Review questions

1 **a**

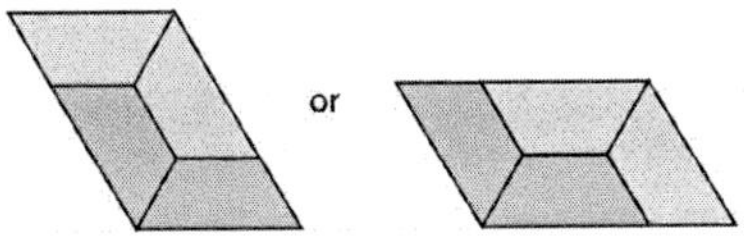

b

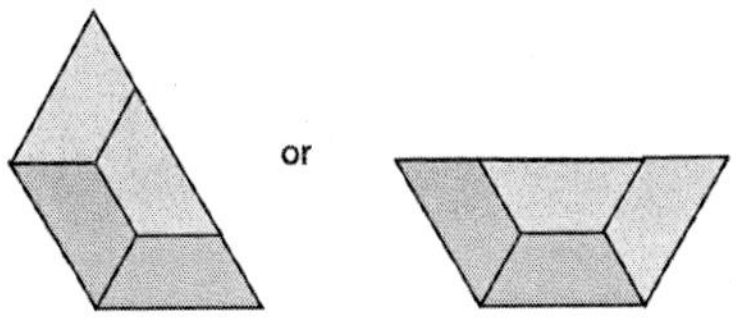

2 **a**

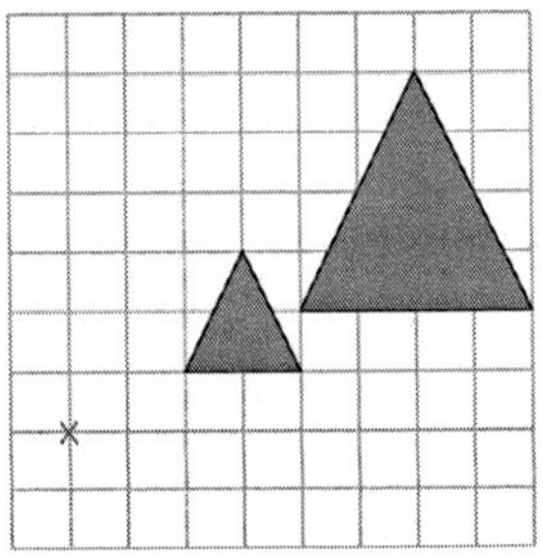

b

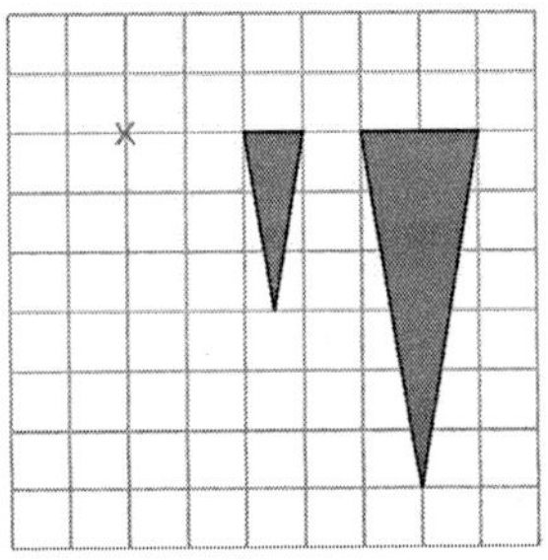

3

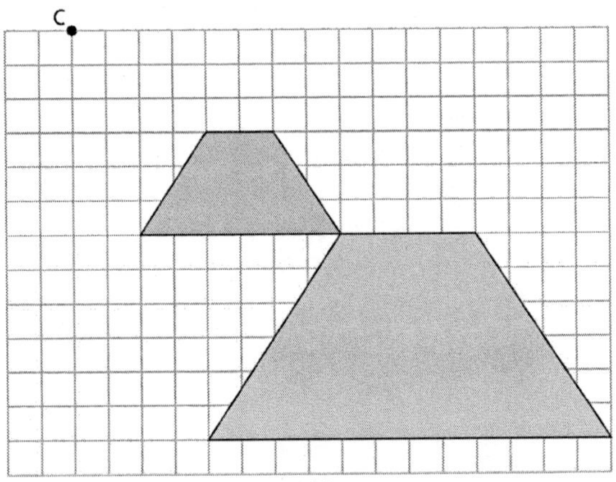

4 **a** 13:25 **b** 48%

5 **a** 210 km **b** 225 km **c** 141 km **d** 105 km

6 **a** $4(x + 3) + 10(x + 3) + 20 = 118$, $4x + 12 + 10x + 30 + 20 = 118$, $14x + 62 = 118$

b $x = 4$ **c** 70 cm^3 **d** 80.5 cm^3

7 A and C, three equal sides (SSS)

Answers to Problem solving – Photographs

1 £183.96

2 **a** 5.1” × 3.5” and 17.7” × 11.8”

b 6 sq in, 24 sq in, 35 sq in, 48 sq in, 80 sq in, 96 sq in

c 6” × 4” and 8” × 6”, 8” × 6” and 12” × 8”

3 **a** 13 sq in **b** Frame B – 6” × 4”, Frame C – 10” × 8”

4 3” × 2” and 6” × 4” SF = 2, 6” × 4” and 12” × 8” SF = 2, 3” × 2” and 12” × 8” SF = 4

5 **a** 6” × 4”, 7” × 5” and 8” × 6” **b** FastPrint, £3.60 **c** 12p **d** 10%

6 **a** 3 : 2 = 1.5 : 1, 6 : 4 = 1.5 : 1, 7 : 5 = 1.4 : 1, 8 : 6 = 1.33 : 1, 10 : 8 = 1.25 : 1, 12 : 8 = 1.5 : 1

b 3” × 2”, 6”× 4” and 12”× 8”

12 Fractions and decimals

Learning objectives

- How to multiply fractions and integers
- How to divide fractions and integers
- How to multiply large and small numbers
- How to divide large and small numbers

Prior knowledge

- The relationship between mixed numbers and improper fractions
- How to add two fractions or mixed numbers
- How to subtract two fractions or mixed numbers
- How to find the lowest common multiple (LCM)

Context

- This chapter will help pupils to understand the relationship between decimals and fractions as being *different representations of parts of a whole*. In modern society we use fractions, decimals and percentages as ways to present parts of a whole. A decimal fraction is a *fraction* whose *denominator* is a *power* of ten. We use decimal notation to write these decimal fractions using a decimal point and zeros as place holders. For example, $\frac{7}{10}$, $\frac{72}{100}$, $\frac{72}{1000}$ and $\frac{72}{10000}$ are written as 0.7, 0.72, 0.072, and 0.0007 respectively. The *whole part*, or integral part of a decimal number is the part to the left of the decimal point. The part to the right of the decimal point is the *fractional part*.

Discussion points

- Why do 25 ÷ 10 and 250 ÷ 100 give the same answer?
- My calculator display shows 0.001. Tell me what will happen when I multiply by 100. What will the display show?
- I divide a number by 10, and then again by 10. The answer is 0.3. What number did I start with? How do you know?
- How would you explain how to multiply a decimal by 10, and how to divide a decimal by 100?

Associated Collins ICT resources

- Chapter 12 interactive activities on Collins Connect online platform
- *Multiplying decimals* video on Collins Connect online platform
- *Mapping the Globe* and *Art of decimals* Wonder of Maths on Collins Connect online platform

Curriculum references

Develop fluency

- Consolidate their numerical and mathematical capability from key stage 2 and extend their understanding of the number system and place value to include decimals, fractions, powers and roots
- Move freely between different numerical representations [for example, equivalent fractions]

Solve problems

- Develop their mathematical knowledge, in part through solving problems and evaluating the outcomes, including multi-step problems

Number

- Order positive and negative integers, decimals and fractions; use the number line as a model for ordering of the real numbers; use the symbols $=$, $\neq$, $<$, $>$, $\leq$, $\geq$ Use the four operations, including formal written methods, applied to proper and improper fractions and mixed numbers

Fast-track for classes following a 2-year scheme of work

- Pupils should be familiar with how to use all four operations with fractions. However, these operations are important concepts and pupils often get them confused. While working through the examples in Lessons 12.1 to 12.3, check pupils' confidence with using the examples. If pupils demonstrate confidence, then concentrate on the challenge activities at the end of each exercise in these lessons. Then move on to Lesson 12.4 and Lesson 12.5.

Lesson 12.1 Adding and subtracting fractions

Learning objective

- To add and subtract fractions and mixed numbers

Links to other subjects

- **Physical Education** – to work out results for sporting activities
- **Art** – to calculate the layout of a painting such as using the rule of thirds

Resources and homework

- Pupil Book 2.2, pages 209–211
- Homework Book 2, section 12.1
- Online homework 12.1, questions 1–10

Key words

- No new key words for this topic

Problem solving and reasoning help

- **MR** question 10 of Exercise 12A in the Pupil Book makes links across different areas of mathematics; in this case, sequences. The question also encourages pupils to hypothesise and then check their answers. The challenge at the end of the exercise introduces a fractional magic square. Pupils will probably have worked on magic squares before so they can apply what they know to this new context. Before tackling the challenge question, you may want to revisit the standard magic square. Encourage pupils to explain how they identify the magic number.

Common misconceptions and remediation

- Pupils often get taught rules without fully understanding them. As a result, pupils may struggle to apply the rules in different contexts and often confuse the rules for adding and subtracting fractions with those for multiplying and dividing fractions. Another common problem is that pupils add or subtract the denominators as well as the numerators when adding and subtracting fractions.

Probing questions

- Why are equivalent fractions important when adding or subtracting fractions?
- What strategies do you use to find a common denominator when adding or subtracting fractions? Is there only one possible common denominator? What happens if you use a different common denominator?
- Give pupils some examples of addition and subtraction of fractions with common mistakes in them. Ask them to talk you through the mistakes and how they would correct them.

Part 1

- Use a number line drawn on the board, or a 'counting stick'. Mark or state that one end is the number 0 and the other end is the number 4, as shown:

- Ask pupils to identify the other marks on the stick.
- As a group, or with an individual, count in units of 0.4.
- Point to the positions on the line or stick until you reach the end of it. Then let pupils continue with no prompts. Repeat, possibly with different individuals to establish a class 'record'.
- Repeat the activity with the same line or stick, but marked with 0 and 7 at either end. Alternatively, count down from 4 or 7 to 0.

Part 2

- Write the following problem on the board: $\frac{1}{2} + \frac{1}{3}$.
- Ask pupils to say or write down the answer. Most will say: $\frac{2}{5}$. Once someone has given the correct answer, discuss the processes used to add fractions with a different denominator.
- Pupils have met this before and should recall the need to use the same denominator.
- Work through the original example in some detail: $\frac{1}{2} + \frac{1}{3} = \frac{3}{6} + \frac{2}{6} = \frac{5}{6}$.
- Repeat with: $\frac{3}{8} + \frac{5}{12} = \frac{9}{24} + \frac{10}{24} = \frac{19}{24}$, and: $\frac{4}{5} + \frac{3}{4} = \frac{16}{20} + \frac{15}{20} = \frac{31}{20} = 1\frac{11}{20}$.
- Recall the need to write top-heavy fractions as mixed numbers.
- Now ask for the answer to: $\frac{2}{3} - \frac{1}{4} = \frac{8}{12} - \frac{3}{12} = \frac{5}{12}$.
- The method is the same, but with the top two numbers subtracted rather than added.
- Repeat with: $\frac{4}{5} - \frac{2}{3} = \frac{12}{15} - \frac{10}{15} = \frac{2}{15}$, and $\frac{5}{12} - \frac{3}{8} = \frac{10}{24} - \frac{9}{24} = \frac{1}{24}$.
- **Pupils can now do Exercise 12A from Pupil Book 2.2.**

Part 3

- Write some fractions on the board, including some mixed numbers (for example, $\frac{1}{7}$, $\frac{3}{8}$, $\frac{3}{14}$, $\frac{5}{6}$, $2\frac{1}{3}$, $1\frac{2}{5}$). Ask pupils to add and subtract combinations of these.
- Make sure pupils are aware of the need to convert mixed numbers to top-heavy fractions or to do the integer part separately.

Answers

Exercise 12A

1	**a** $\frac{7}{8}$	**b** $1\frac{1}{6}$	**c** $1\frac{1}{2}$	**d** $1\frac{1}{2}$	**e** $1\frac{5}{12}$	**f** $1\frac{1}{10}$	**g** $1\frac{5}{24}$	**h** $1\frac{1}{12}$
2	**a** $1\frac{5}{8}$	**b** $2\frac{5}{8}$	**c** $2\frac{5}{6}$	**d** $3\frac{11}{20}$	**e** $4\frac{1}{12}$	**f** $5\frac{17}{24}$	**g** $13\frac{5}{6}$	**h** $5\frac{5}{24}$
3	**a** $\frac{1}{2}$	**b** $\frac{1}{4}$	**c** $\frac{5}{8}$	**d** $\frac{9}{20}$	**e** $\frac{11}{24}$	**f** $\frac{7}{15}$	**g** $\frac{13}{20}$	**h** $\frac{7}{30}$
4	**a** $1\frac{5}{8}$	**b** $1\frac{1}{2}$	**c** $1\frac{5}{8}$	**d** $\frac{5}{8}$	**e** $1\frac{3}{4}$	**f** $2\frac{1}{6}$	**g** $2\frac{1}{3}$	**h** $2\frac{7}{10}$
5	**a** $\frac{1}{4}$	**b** $1\frac{1}{3}$	**c** $\frac{7}{8}$	**d** $1\frac{3}{8}$				
6	**a** $1\frac{1}{2}$	**b** $1\frac{3}{4}$	**c** $3\frac{3}{10}$	**d** $2\frac{3}{8}$				
7	**a** $2\frac{5}{8}$	**b** $2\frac{3}{4}$	**c** $3\frac{1}{8}$	**d** $5\frac{4}{5}$				

8 **a** $13\frac{1}{2}$ cm **b** $10\frac{1}{4}$ cm **c** 9 cm

9 $7\frac{1}{6}$ cm

10 **a i** $\frac{1}{4}$ **ii** $\frac{3}{8}$ **iii** $\frac{5}{16}$ **b i** $\frac{11}{32}$ **ii** Pupils check own answer

11 **a** $\frac{3}{8}$ **b** $\frac{1}{24}$

Challenge: Magic square

A 1 **B** The three rows are: $\frac{2}{15}$, $\frac{3}{5}$, $\frac{4}{15}$; $\frac{7}{15}$, $\frac{1}{3}$, $\frac{1}{5}$; $\frac{2}{5}$, $\frac{1}{15}$, $\frac{8}{15}$

Lesson 12.2 Multiplying fractions and integers

Learning objective

- To multiply a fraction and an integer

Resources and homework

- Pupil Book 2.2, pages 211–214
- Homework Book 2, section 12.2
- Online homework 12.2, questions 1–10

Links to other subjects

- **Physical Education** – to work out results for sporting activities
- **Geography** – to work out distances and/or areas from a map

Key words

- No new key words for this topic

Problem solving and reasoning help

- The challenge activity at the end of Exercise 12B in the Pupil Book requires pupils to apply their understanding to a variation on the classic multiplication table.

Common misconceptions and remediation

- Pupils often get taught rules without fully understanding them, so they may struggle to apply the rules in different contexts. Pupils often confuse the rules for multiplying and dividing fractions with the rules for adding and subtracting fractions.

Probing questions

- 'Multiplying makes numbers bigger.' When is this statement true and when is it false?
- 18 (x 0.5. Talk me through how you would justify the answer to this calculation to someone who does not know the rules for multiplying a whole number by a decimal?

Part 1

- Display a target board as shown:

10	15	20	25	30	35
6	12	18	24	36	42
22	48	56	80	72	100

- Ask pupils to give, for example, halves of the numbers in the bottom row, thirds of the numbers in the middle row, and fifths of the numbers in the top row.
- Now ask pupils to give two-thirds of the numbers in the middle row. Discuss how to do this by working out one-third and then doubling.
- Extend to other non-unit fractions, for example, two-fifths.

Part 2

- Following on from the starter, reinforce that, for example, $\frac{2}{5}$ of 30 is the same as $\frac{2}{5} \times 30$.
- Encourage pupils to use mental arithmetic rather than a calculator.
- Now ask for the answer to $\frac{4}{5}$ of 45.
- Recall the method to do this: find $\frac{1}{5}$ of 45 (= 9) and then multiply by 4 (= 36).
- Repeat with $\frac{3}{4}$ of £28 (= £21), $\frac{2}{7}$ of 35 kg (= 10 kg).

- Ask for the answers to $3 \times \frac{4}{7}$ (= $\frac{12}{7}$ = $1\frac{5}{7}$) and $5 \times \frac{4}{5}$ (= $\frac{20}{5}$ = 4).
- Use further examples if pupils need more practice.
- **Pupils can now do Exercise 4B from Pupil Book 2.2.**

Part 3

- To consolidate the learning from this lesson, ask pupils to work in pairs to answer this: 'In a survey of 24 pupils, $\frac{1}{3}$ liked football best, $\frac{1}{4}$ one-quarter liked basketball, $\frac{3}{8}$ liked athletics and the rest liked swimming. How many pupils liked swimming?'

Answers

Exercise 12B

1	**a** $2\frac{1}{3}$	**b** $3\frac{3}{5}$	**c** $1\frac{3}{4}$	**d** $2\frac{2}{3}$				
2	**a** 3	**b** 4	**c** $2\frac{1}{2}$	**d** $\frac{3}{8}$	**e** $1\frac{3}{4}$	**f** $1\frac{4}{5}$	**g** $1\frac{1}{4}$	**h** $\frac{1}{2}$
3	**a** $1\frac{1}{3}$	**b** $2\frac{2}{5}$	**c** $2\frac{1}{4}$	**d** $1\frac{3}{4}$	**e** 3	**f** $3\frac{1}{5}$	**g** $2\frac{3}{4}$	**h** $2\frac{1}{4}$
4	**a** $3\frac{3}{4}$	**b** $6\frac{1}{4}$	**c** $3\frac{1}{5}$	**d** $8\frac{1}{2}$	**e** 11	**f** $8\frac{2}{5}$	**g** $6\frac{1}{2}$	**h** $4\frac{1}{4}$
5	**a** $12\frac{1}{2}$ cm²		**b** $31\frac{1}{4}$ cm²		**c** $8\frac{1}{4}$ cm²			
6	**a** $2\frac{3}{4}$	**b** $21\frac{1}{4}$ cm²						
7	**a** $38\frac{1}{2}$	**b** $49\frac{1}{2}$	**c** $31\frac{1}{5}$	**d** $92\frac{1}{2}$	**e** $102\frac{1}{2}$	**f** $101\frac{1}{3}$	**g** $166\frac{2}{3}$	**h** $80\frac{7}{16}$
8	**a** 8	**b** $11\frac{1}{3}$	**c** $20\frac{2}{3}$					
9	**a** $1\frac{1}{8}$	**b** $4\frac{1}{2}$	**c** $11\frac{1}{4}$					
10	**a** $3\frac{3}{4}$	**b** $6\frac{7}{8}$	**c** $17\frac{1}{2}$					

Challenge: Multiplication table

Top row: $4\frac{1}{2}$, $7\frac{1}{2}$, $10\frac{1}{2}$, 15

Second row: $6\frac{3}{4}$, $11\frac{1}{4}$, $15\frac{3}{4}$, $22\frac{1}{2}$

Third row: 11, $18\frac{1}{3}$, $25\frac{2}{3}$, $36\frac{2}{3}$

Fourth row: $12\frac{1}{2}$, $20\frac{5}{6}$, $29\frac{1}{6}$, $41\frac{2}{3}$

Lesson 12.3 Dividing with integers and fractions

Learning objectives

- To divide a fraction or a mixed number by an integer
- To divide an integer by a unit fraction

Links to other subjects

- **Physical Education** – to work out results for sporting activities
- **Geography** – to work out distance and/or areas from a map

Resources and homework

- Pupil Book 2.2, pages 214–216
- Homework Book 2, section 12.3
- Online homework 12.3, questions 1–10

Key word

- unit fraction

Problem solving and reasoning help

- In the challenge activity at the end of Exercise 12C of the Pupil Book, pupils are required to apply their learning to a sporting context. Ask pupils to translate this to sporting activities of their own.

Common misconceptions and remediation

- Pupils often get taught rules without fully understanding them, so pupils may struggle to apply the rules in different contexts. Pupils often confuse the rules for multiplying and dividing fractions with those for adding and subtracting fractions.

Probing questions

- How would you justify that dividing by $\frac{1}{2}$ is the same as multiplying by 2? What about dividing by $\frac{1}{3}$ and multiplying by 3? What about dividing by $\frac{2}{3}$ and multiplying by $\frac{3}{2}$?
- $0.6 \div 0.1$, $2.4 \div 0.5$, $2.4 \div 0.2$. Explain the reasoning that led you to the answer to each of these calculations.

Part 1

- Working in pairs, ask pupils to consider the following statement: 'Division makes things smaller.' When is this statement true and when is it false?
- Take feedback asking pupils to justify their response.

Part 2

- Explain to pupils that in this lesson they will use division with a unit fraction and an integer. Remind pupils that a unit fraction is a fraction with a numerator of 1.
- Give two examples such as: $\frac{1}{2} \div 5$ and $5 \div \frac{1}{2}$.
- Say to pupils that they can think of $\frac{1}{2} \div 5$ as meaning one-fifth of $\frac{1}{2}$.
- The answer is $\frac{1}{10}$ because $\frac{1}{10} \times 5 = \frac{1}{2}$
- Tell pupils that they can also write $\frac{1}{2} \div 5$ as $\frac{1}{2 \times 5} = \frac{1}{10}$. Just multiply the denominator by 5.
- Pupils can also think of $5 \div \frac{1}{2}$ as: 'Work out how many halves make 5.'
- What do they think the answer to this might be? (Answer is 10 because $\frac{1}{2} \times 10 = 5$)
- You can write: $5 \div \frac{1}{2} = 10$. Just multiply 5 by the denominator.

- To help clarify this, pupils can work in pairs through Example 5 on page 215 of the Pupil Book. Encourage pupils to ask any questions they may still have.
- You could use visual images to help pupils who struggle.
- Pupils can now do Exercise 12C from Pupil Book 2.2.

Part 3

- To clarify and extend their understanding, you could ask pupils to respond to this question: 'How would you justify that $4 \div \frac{1}{5} = 20$? How would you use this to work out $4 \div \frac{2}{5}$?
  Do you expect the answer to be greater or less than 20? Why?
- You could approach this question as a class discussion using the stems, 'I agree or disagree with ... because ...'.

Answers

Exercise 12C

1 **a** $\frac{1}{4}$ **b** $\frac{1}{8}$ **c** $\frac{1}{12}$ **d** $\frac{1}{20}$

2 **a** $\frac{1}{3}$ **b** $\frac{2}{9}$ **c** $\frac{1}{6}$ **d** $\frac{1}{9}$

3 **a** $\frac{3}{16}$ **b** $\frac{3}{20}$ **c** $\frac{5}{48}$ **d** $\frac{1}{24}$ **e** $\frac{7}{16}$ **f** $\frac{2}{7}$ **g** $\frac{1}{18}$ **h** $\frac{1}{10}$

4 **a** $\frac{3}{4}$ **b** $1\frac{1}{4}$ **c** $1\frac{3}{4}$ **d** $1\frac{5}{8}$

5 **a** $\frac{5}{8}$ **b** $\frac{7}{9}$ **c** $1\frac{1}{2}$ **d** $1\frac{3}{8}$ **e** $\frac{7}{10}$ **f** $1\frac{5}{8}$ **g** $1\frac{7}{10}$ **h** $\frac{11}{12}$

6 **a** $1\frac{7}{8}$ **b** $1\frac{1}{2}$ **c** $\frac{15}{16}$ **d** $\frac{3}{4}$

7 $3\frac{3}{8}$ cm

8 $3\frac{7}{12}$ cm

9 **a** 8 **b** 10 **c** 9 **d** 8

10 **a** 16 **b** 24 **c** 40 **d** 64

11 **a** 20 **b** 20 **c** 20 **d** 20

12 $\frac{1}{3}$

13 90 seconds

Challenge: Cycle race

A **a** 15 **b** 20

B **a** 168 **b** 224

Lesson 12.4 Multiplication with large and small numbers

Learning objectives

- To multiply with combinations of large and small numbers mentally

Resources and homework

- Pupil Book 2.2, pages 216–219
- Intervention Workbook 2, pages 20–21
- Homework Book 2, section 12.4
- Online homework 12.4, questions 1–10

Links to other subjects

- **Science** – to using real-life examples in biology and chemistry (see STEM nrich website for examples)

Key words

- No new key words for this topic

Problem solving and reasoning help

- **PS** question 12 of Exercise 12D in the Pupil Book and the challenge activity at the end of the exercise give pupils the opportunity to see how their understanding of decimals can be crucial across all areas of mathematics. In this case, volume and completing graphs and tables are used as examples.

Common misconceptions and remediation

- Pupils often find decimal fractions difficult to understand. They may have encountered them, without realising, when solving money problems. Working with decimals is an extension of pupils' understanding of place value. Most pupils have an understanding that each column to the left of another is 10 times greater. Build on this so that pupils are aware that each column to the right is 10 times smaller.

Probing questions

- Explain how you would do this multiplication using factors such as: 4.7×60.
- Give pupils some examples of division with large and small numbers with mistakes in them. Ask pupils to look for the mistakes. Then ask pupils to provide feedback on what the mistakes are and how these might be corrected.

Part 1

- This activity has to do with adding and subtracting 0.1, 0.01 and 0.001 from other decimals and whole numbers. Ask pupils to use mini whiteboards to write their answers.
- Give the class an example such as, 'What number is 0.01 more than 6.03?' (= 6.04)
- Now give more examples and ask pupils to hold up their answers on their white boards.
- Suggested questions are:
  - What is 0.001 more than 2.008? (= 2.009)
  - What is 0.01 less than 5? (= 4.99)
  - What is 0.002 less than 3? (= 2.998)
  - What needs to be added to 3.234 to make 3.237? (= 0.003)
  - What needs to be subtracted from 4.002 to make 3.997? (= 0.005)

Part 2

- Refer to the introduction to this lesson on page 216 of the Pupil Book. Use it as a class activity to prepare for this lesson. For example, all the answers can be deduced from the simple multiplication 3 × 2 = 6. The problem is where to put the decimal point.
    - 300 × 2 = 600
    - 0.3 × 20 = 6
    - 0.03 × 0.2 = 0.006
    - 30 × 0.002 = 0.06
- Take feedback encouraging pupils explain *how* they arrive at their answers by comparison with the simple multiplication 3 × 2 = 6.
- Then work through Example 8 on page 217 to help pupils see how they can use their understanding of place value to help them to divide large and small numbers.
- **Pupils can now do Exercise 12D from Pupil Book 2.2.**

Part 3

- Write a variety of large numbers on the board such as:

  4 502 611   5 560 097   2 110 009   7 899 911
- Ask pupils to read out the numbers and round them off to the nearest ten thousand, hundred thousand and million.

Answers

Exercise 12D

1	**a** 1200	**b** 1200	**c** 12 000	**d** 120 000
2	**a** 4000	**b** 4200	**c** 72 000	**d** 72 000
3	**a** 3.2	**b** 0.09	**c** 0 035	**d** 0.45
	e 0.008	**f** 0.036	**g** 0.0072	**h** 0.066
4	**a** 400	**b** 1600	**c** 3600	**d** 90 000
	e 0.09	**f** 0.81	**g** 0.0004	**h** 0.0064
5	**a** 600 cm²	**b** 200 mm and 300 mm		**c** 60 000 mm²
	d 0.2 m and 0.3 m		**e** 0.06 m²	
6	**a** 1.8	**b** 24	**c** 27	**d** 200
	e 64	**f** 6.6	**g** 630	**h** 1.5
7	**a** 12	**b** 120	**c** 1.2	**d** 12
8	**a** 4800 cm²	**b** 0.48 m²		
9	**a** 63	**b** 6.3	**c** 63 000	**d** 0.063
10	**a** 52 900	**b** 5.29	**c** 0.0529	
11	**a** 24	**b** 240	**c** 2400	**d** 0.24
12	**a** 0.16 m³	**b** 160 000 cm³		

Challenge: Falling down

A The missing values are 0.05, 0.2, 0.8, 1.25.

B Check pupils' graphs.

Lesson 12.5 Division with large and small numbers

Learning objective

- To divide combinations of large and small numbers mentally

Resources and homework

- Pupil Book 2.2, pages 219–221
- Homework Book 2, section 12.5
- Online homework 12.5, questions 1–10

Links to other subjects

- **Science** – to use real-life examples in biology and chemistry (see STEM nrich website for examples)

Key word

- divisor

Problem solving and reasoning help

- The reasoning activity at the end of Exercise 12E in the Pupil Book gives pupils the chance to makes links between decimals and fractions. Encourage pupils to discuss how, when making a choice between using decimals and fractions, how they might make a decision.

Common misconceptions and remediation

- Pupils often find decimal fractions difficult to understand. They may have encountered them without realising when solving money problems. Working with decimals is an extension of pupils' understanding of place value. Most pupils understand that each column to the left of another is 10 times greater. Build on this so that pupils are aware that each column to the right is 10 times smaller.

Probing questions

- 73.6 ÷ 3.2 = 23. Explain how you can use this to devise calculations with the same answer.

Part 1

- Ask pupils to work in pairs to answer the following question.
  37 × 64 = 2368. 'Explain how you can use this fact to devise calculations with answers of: 23.68, 2.368, 0.2368.'
- Encourage pupils to explain *how* they arrived at their solutions.

Part 2

- Use the introduction to this lesson on page 219 of the Pupil Book, as follows: Ask pupils to work in pairs and to look at these divisions. They all have the same answer.
  - 8 ÷ 4 = 2
  - 80 ÷ 40 = 2
  - 800 ÷ 400 = 2
  - 0.8 ÷ 0.4 = 2
  - 0.08 ÷ 0.04 = 2
- Take feedback, asking pupils to explain why each answer is the same.
- Ask pupils how they could use multiplication to check that the answers are correct.
  For example, 2 × 40 = 80 and 2 × 0.04 = 0.08.
- Then work through the Example 9 on page 220. This will help pupils to see how they can use their understanding of place value to help them divide large and small numbers.
- **Pupils can now do Exercise 12E from Pupil Book 2.2.**

Part 3

- Give pupils some examples of division with large and small numbers with mistakes in them.
- Ask pupils to identify the mistakes and ask them to provide feedback on what the mistakes are and how they might be corrected.

Answers

Exercise 12E

1 **a** 30 **b** 20 **c** 30 **d** 30
e 50 **f** 30 **g** 40 **h** 50

2 **a** 150 **b** 50 **c** 200 **d** 200
e 2000 **f** 300 **g** 500 **h** 200

3 **a** 20 **b** 2 **c** 0.2 **d** 0.02

4 **a** 7 **b** 0.7 **c** 0.07 **d** 0.07

5 **a** 0.4 **b** 0.2 **c** 0.7 **d** 0.03
e 0.2 **f** 0.04 **g** 0.3 **h** 0.05

6 **a** 0.06 **b** 0.03 **c** 0.02 **d** 0.01

7 **b** 270 ÷ 90 = 3, the rest are 30

8 **d** 0.8 ÷ 0.2 = 4, the rest are 0.4

9 **a** 30 **b** 0.03 **c** 0.3 **d** 3000

10 **a** 30 **b** 0.3 **c** 0.03 **d** 3; order is c, b, d, a

11 **a** 40 **b** 5 **c** 50 **d** 0.006
e 30 **f** 40 **g** 20 **h** 0.05

12 **a** Changing the subject of the formula $A = 20x$ gives $x = \frac{A}{20}$.
b 0.02 **c** 50

13 **a** 1.3 **b** 0.13 **c** 1.3 **d** 130

Reasoning: Fraction or decimal?

A **a** Answer given in Pupil Book
b 40 **c** 320 **d** 120 **e** 15 **f** 70

B Check pupils' answers

Review questions (Pupil Book pages 222–223)

- The review questions will help to determine pupils' abilities with regard to the material within Chapter 12.
- These questions also draw on the maths covered in earlier chapters of the book to encourage pupils to make links between different topics.
- The answers are on the next page.

Challenge – Guesstimates (Pupil Book pages 224–225)

- This activity gives pupils the opportunity to practise their mental strategies in some real-life contexts. It also encourages pupils to make links to the use of estimation as well as the need to make assumptions when tackling real-life problems.
- Pupils need to realise that they need to have calculations that give them a reasonable approximation but NOT an exact answer.
- Ask pupils further questions so that they think about what an estimate means, for example:
    - Can you think of some example in your life when you might want to make an estimate?
    - How do you decide what a reasonable estimate might be?
    - Does this change depending on what you are going to use the estimation for?
    - What assumptions might you need to make when estimating?
- You may want to model your response to one of the examples, or an alternative example, with pupils. Vocalise and discuss with pupils the types of questions you might ask yourself internally when making the estimate.
- More pupils could explore the difference between rounding before and after a calculation.
- Ask pupils to work through questions 1 to 9 in the activity.

Answers to Review questions

1	a $3\frac{1}{12}$	b $2\frac{7}{12}$	c $\frac{17}{24}$	
2	a $2\frac{2}{3}$	b $3\frac{3}{4}$	c $3\frac{4}{5}$	
3	a $21\frac{1}{3}$ cm	b $26\frac{2}{3}$ cm²		
4	a $\frac{7}{10}$	b $4\frac{1}{5}$		
5	a $1\frac{1}{5}$	b $7\frac{1}{5}$	c $14\frac{2}{5}$	d $12\frac{2}{5}$
6	a 4900	b 100 000	c 90 000	d 2400
7	a 20	b 0.064	c 0.06	d 81
8	a 1400 cm²	b 0.14 m²		
9	a 30	b 80	c 40	d 400
10	a 30	b 40	c 0.4	d 0.06
11	a 288 000	b 28.8		
12	$1\frac{2}{3}$			
13	a $2\frac{1}{5}$	b $\frac{11}{15}$	c $5\frac{8}{15}$	
14	a 75	b 42	c 4.2	d 7.5

Answers to Challenge – Guesstimates

1 5000, 0.2, 5000 ÷ 0.2 = 25 000

2-9 Answers will vary. Check that they are sensible and backed up with working.

13 Proportion

Learning objectives

- How to solve problems involving direct proportion
- Graphical and algebraic representations of direct proportion
- How to solve problems involving inverse proportion
- Graphical and algebraic representations of inverse proportion

Prior knowledge

- How to write a simple formula algebraically
- What a ratio is
- How to draw graphs
- In KS2 pupils were taught how to solve problems involving the relative sizes of two quantities where missing values can be found by using integer multiplication and division facts. This chapter builds further on this and the elements of proportion taught in Year 7.

Context

- This chapter introduces the concepts of direct and inverse proportion as a means of solving practical questions. Section 13.2 concerns graphs showing direct proportion.

Discussion points

- Five miles is about the same as eight km.
- Make up some questions to convert miles to kilometres and vice versa that you could do in your head.
- How would you convert £ to $ and $ to £ using a graph?

Associated Collins ICT resources

- Chapter 13 interactive activities on Collins Connect online platform
- *Understanding inverse proportion* video on Collins Connect online platform
- *Usain Bolt* Wonder of Maths on Collins Connect online platform

Curriculum references

Reason mathematically

- Extend and formalise their knowledge of ratio and proportion in working with measures and geometry, and in formulating proportional relations algebraically.

Solve problems

- Develop their mathematical knowledge, in part through solving problems and evaluating the outcomes, including multi-step problems.

Ratio, proportion and rates of change

- Solve problems involving direct and inverse proportion, including graphical and algebraic representations.

Fast-track for classes following a 2-year scheme of work

- Much of the material in this chapter will be unfamiliar to pupils. Make sure that all pupils understand each concept fully before moving on to the **MR** and **PS** questions in the exercises.

Lesson 13.1 Direct proportion

Learning objectives

- To understand the meaning of direct proportion
- To find missing values in problems involving proportion

Resources and homework

- Pupil Book 2.2, pages 227–230
- Homework Book 2, section 13.1
- Online homework 13.1, questions 1–10

Links to other subjects

- **Art** – to use proportion in paintings and drawings
- **Food technology** – to adjust ingredients in a recipe

Key words

- direct proportion
- proportional
- variable

Problem solving and reasoning help

- The **FS** and **MR** questions in Exercise 13A in the Pupil Book require pupils to apply their understanding of direct proportion and ratio in practical situations.

Common misconceptions and remediation

- Pupils often mix up direct and inverse proportion, usually by using direct proportion to answer inverse proportion questions. Pupils sometimes use an incorrect multiplier.

Probing questions

- Six apples cost 90p. Explain how you would work out how much eight apples or 12 apples cost?
- Three books cost £18. Explain how you would work out how much five books cost?

Part 1

- Give pupils a ratio, such as 2 : 5, and ask them to find the missing value in the equivalent ratio … : 15 (6 : 15). Repeat with 10 : … (10 : 25), … : 35 (14 : 35), and so on.
- Pupils could use number fans or mini whiteboards for writing down their answers.

Part 2

- Introduce the class to the concept of *direct proportion* by giving them an example: 'When a shopper purchases a number of equally-priced articles, the total cost is proportional to the number of articles that the shopper bought. So, if the shopper bought 20 articles, the total cost would be 20 times the cost of one.'
- Problems involving direct proportion are normally solved by referring to a single item. This approach is known as the *unitary method.*
- Now ask pupils to solve the following problem: 'If five exercise books have a total of 200 pages, how many pages do four exercise books have?'
- Pupils may have an intuitive idea of the answer (160 pages) but outline the *unitary method.*
- For example, five books have 200 pages, so one book has 40 pages. Hence, four books have 4 × 40 = 160 pages. Alternatively, consider this as a ratio: 5 : 200 = 1: 40 = 4 : 160.
- Four canteen tables can seat 24 people. How many people can sit at six canteen tables?
- If four tables seat 24 people, one table would seat six people. Therefore, six tables would seat 6 × 6 = 36 people. Alternatively, as a ratio: 4 : 24 = 1 : 6 = 6 : 36.
- Nine pencils cost 81p. How much would 12 pencils cost? One pencil costs 9p, so 12 pencils cost 12 × 9p = £1.08. Alternatively, as a ratio: 9 : 81 = 1 : 9 = 12 : 108.

A bus with 20 passengers takes 10 minutes to travel from Silkstone to Barnsley. How long will the same bus with 10 passengers take to do the same journey? The number of passengers does not affect the speed of the bus, so the time taken will be the same! (10 minutes)

- **Pupils can now do Exercise 13A from Pupil Book 2.2**

Part 3

- First establish that no pupils were having difficulties with Exercise 13A. If no pupils were, then ask them to consider the following problem.
- Six men take four days to lay a pipeline. How long would it take three men to lay the same pipeline?
- Pupils may well give the intuitive answer of two days. So, explain that it will take fewer men more time to do the job. Half the men will take twice the time (eight days).
- Then, ask how long two men would take to lay the pipeline? (12 days)
- Next, ask how long one man would take to lay the pipeline? (24 days)
- Ask pupils if they can see a connection between the answers (product is 24).
- This is an example of inverse proportion, which we will meet later in this chapter.

Answers

Exercise 13A

1 Missing numbers are 26, 52, 78 and 117
2 **a** £1.68 **b** £2.52 **c** 42p **d** 28p
3 **a** 54 g **b** 108 g **c** 216 g **d** 13.5 g
4 **a** 6 kg **b** 24
5 15, 25 and 125 on the top row, 64 and 160 on the bottom row
6 **a** 22 m **b** 33 m **c** 110 m **d** 5.5 m
7 **a** 320 **b** $2\frac{1}{2}$ hours
8 **a i** £1.92 **ii** £3.20 **iii** 16p
b i 200 g **ii** 1 kg **iii** 50 g
9 82 and 410 on the top row, 315 and 504 on the bottom row
10 0.7 and 1.4 on the top row, 120 and 180 on the bottom row
11 **a** 32.8 kg **b** 30 m
12 **a** 288 **b** 1 : 3 **c** 1 : 3
13 **a** 190 **b** Both are 1 : 5
14 No. A possible explanation is $50 \times 2 = 100$ but $122 \times 2 \neq 212$.

Investigation: Age, height and mass

A No. Own explanation
B No. Own explanation
C No. Own explanation

Lesson 13.2 Graphs and direct proportion

Learning objective

- To represent direct proportion graphically and algebraically

Resources and homework

- Pupil Book 2.2, pages 230–234
- Homework Book 2, section 13.2
- Online homework 13.2, questions 1–10

Links to other subjects

- **Business studies** – to use currency conversion graphs
- **Science** – to use conversion graphs

Key words

- formula
- graph

Problem solving and reasoning help

- The 'Financial skills: Exchange rates' activity at the end of Exercise 13B of the Pupil Book requires the class to apply their learning to a currency conversion graph and to find a formula to convert between the two currencies. Revise how to work out the equation of a straight line with pupils before they start this activity.

Common misconceptions and remediation

- Some pupils may struggle to draw correctly-scaled axes for the questions in this exercise. Draw the scales for these pupils and explain the reasoning behind the chosen values.

Probing questions

- Give pupils the fact £1 = $1.5 or something similar. Say: Explain how you would use this information to draw a currency conversion graph to convert up to £100 into $. What scales would you use?

Part 1

- Tell the class that you were looking at a DVD yesterday and noted that the running time was 135 minutes. Ask pupils how many hours and minutes this is (2 hours 15 minutes).
- Discuss with pupils how they should calculate this. Pupils should use multiples of 60 to find the number of hours; the remainder is minutes.
- Now ask 'What fraction of an hour is 15 minutes?' Ask for both a fraction and a decimal answer. ($\frac{1}{4}$ and 0.25)
- Now ask: 'What fraction of an hour is 10 minutes?' This is $\frac{1}{6}$ as a fraction but the decimal is awkward. Ask: What will the decimal answer be without using a calculator?
- Starting from known facts, $\frac{1}{3} = 0.3333333$; $\frac{1}{6}$ is half of this, so by short division, $\frac{1}{6} = 0.1666667$. Alternatively, $\frac{3}{6} = \frac{1}{2} = 0.5$, and $\frac{4}{6} = \frac{2}{3} = 0.6666667$, so $\frac{1}{6} = \frac{4}{6} - \frac{3}{6} = 0.6666667 - 0.5 = 0.1666667$.
- Ask: 'What fraction of an hour is five minutes?'. This is $\frac{1}{12}$, which is half of $\frac{1}{6}$ or 0.083333.
- Discuss the potential confusion of using decimal notation in time, for example, 1.50 could mean $1\frac{1}{2}$ hours or 1 hour 50 minutes. Make sure that pupils know which unit is being used.

Part 2

- Tell the class that you took a taxi journey late at night and watched the taxi fare change on the meter display. The table shows the fares at various distances after getting into the taxi.

Distance (km)	0	5	15	20
Fare (£)	1.30	3.00	6.40	8.10

- Draw a pair of axes with 'distance' on the *x*-axis and 'fare' on the *y*-axis. Label them. Make sure the *y*-axis goes up to £10 and the *x*-axis goes as far as 20 km. Now plot the points and draw a suitable straight line through them.
- Explain to pupils that as the line is straight, you can say that the two variables have a linear relationship. Furthermore, the fare increase is directly proportional to the distance travelled. This means that the fare always increases by the same amount for a given increase in distance. You can show that this works out as £1.70 for every 5 kilometres.
- Ensure that all pupils understand direct proportion: For every unit that one variable increases the other variable also increases by a fixed amount. In the example above, the distance travelled and the fare increase after the meter reaches £1.30 are in direct proportion.
- Ask pupils what the equation of the line will be in terms of distance, *m*, in kilometres, and fare, *f*, in pounds.
- Remind pupils of the general equation of a straight line: $y = mx + c$, where m is the gradient and c is the *y*-axis intercept.
- In 20 kilometres, the fare has increased by £6.80. This is £0.34 per kilometre. The fixed charge is £1.30, so the equation is f = 1.30 + 0.34m.
- Ask what the fare would be at 35 kilometres if the same rate continued.
- You can change the graph, or use the equation, so giving £13.20.
- **Pupils can now do Exercise 13B from Pupil Book 2.2.**

Part 3

- Ask pupils to explain the term 'directly proportional'.
- Pupils need to understand that the term means that one variable always increases by the same amount for a given increase in the other variable. The graph of two variables that are directly proportional will always be linear (that is, a straight line) and pass through the origin.
- Discuss the fact that it was not always easy to find the equation of the lines in Exercise 13B, and that using the idea of $y = mx + c$ will always help them in this respect.

Answers

Exercise 13B

1 **a** 8, 20, 28, 32, 40 **b** Check pupils' graphs
2 **a** 1.3 × 10 = 13 **b** 13, 26, 32.5, 39, 52 **c** Check pupils' graphs
3 **a** 10, 50, 100 **b** Check by multiplying **c** Check pupils' graphs
4 **a** 24, 72, 144 **b** 48 **c** $y = 48x$ **d** £3.60 **e** Check pupils' graphs
5 **a** 300, 600, 900, 1200 **b** $y = 12x$ **c** HK$15 240
6 **a** 3, 6, 7, 8.5, 11 **b** $y = 0.1x$ **c** 43 litres
7 **a** 36 km/h **b** 3.6 **c** $y = 3.6x$
8 **a** 14, 21, 28 and 35 **b** $y = 1.4x$ **c** **i** 16.8 mm **ii** 26.6 mm **iii** 43.4 mm
9 **a** $y = 2x$ **b** 9.4 and 3.2 on the top, 10.4 and 25.8 on the bottom
c Check pupils' drawings

Financial skills: Exchange rates

A $y = 1.7x$. This follows from choosing a point on the graph, such as (10, 17)
B Multiply number of rupees by 1.7
C Yes, divide number of yen by 1.7

Lesson 13.3 Inverse proportion

Learning objectives
- To understand what inverse proportion is
- To use graphical and algebraic representations of inverse proportion

Resources and homework
- Pupil Book 2.2, pages 234–238
- Homework Book 2, section 13.3
- Online homework 13.3, questions 1–10

Links to other subjects
- **Science** – to use wavelength and frequency graphs from experiments
- **Business studies** – to use financial graphs

Key word
- inverse proportion

Problem solving and reasoning help
- The MR question requires pupils to produce a formula linking the base and height of a triangle. Model some examples of this type of question for pupils before they tackle this. Explain carefully how to get the constant from the given values of the two variables.

Common misconceptions and remediation
- Pupils will often use direct proportion rather than inverse proportion. Make sure that pupils are aware of the keywords in a question that indicate inverse proportion. Also make sure that pupils write the variables the correct way around in their formula as this is often a common error.

Probing questions
- How do you know whether given sets of numbers are in inverse proportion?
- What tips would you give to help someone them solve word problems involving inverse proportion?

Part 1
- Ask pupils the following questions:

  1 What is nine squared? (81)

  2 What is the square root of 144? (12)

  3 The diagram (draw this on the board) shows a cube made from smaller unit cubes. What is the volume of the whole cube? (8 cm^3)

  4 What is eight squared? (64)

  5 What is the square root of 36? (6)

  6 What is two cubed? (8)

  7 A cube has a volume of 27 cm^3. What is the side of the cube? (3 cm)

  8 The length of one side of a square is 4 cm. What is the area of the square? (16 cm^2)

  9 The area of a square is 64 cm^2. What is the perimeter of the square? (32 cm)

  10 x squared is 4. What two values can x have? (2 and –2)
- If time allows, discuss the methods of solution when giving the answers.
- Repeat these questions within a few days to see if pupils get more answers right.

Part 2

- Introduce the concept of inverse proportion by explaining that it occurs when an increase in one quantity leads to a decrease in the other quantity involved. Again, many inverse proportion problems are best solved by the unitary method.
- Pupils find inverse proportion problems quite difficult, and most answers will be intuitive. For example, as one quantity doubles, the other quantity halves.
- Ask pupils to do this problem:
- Two people build a wall in 10 days. How long will one person take? (20 days) How long will five people take? (4 days) How long will 10 people take? (2 days)
- Repeat with the following examples.
- Some counters can be arranged in 12 lines of 18. How many lines would there be if 24 counters were in each line? (12 lots of 18 is the same as 24 lots of 9; there will be 9 rows.)
- Three taps can fill a sink in 15 minutes. How long would five taps take to fill the sink? (One tap will take three times as long to fill the sink, giving 3 × 15 = 45 minutes. Five taps will therefore take one-fifth of the time that one tap takes, giving 45 ÷ 5 = 9 minutes.)
- When I travel at 60 mph my car does 10 miles to a litre of petrol. How many miles will I get per litre if I travel at 30 mph? (This is impossible to answer as the petrol consumption of a car is not dependent just on speed.)
- **Pupils can now do Exercise 13C from Pupil Book 2.2.**

Part 3

- Write the following on the board: $xy = 24$.
- Ask pupils what values x and y could take. Write these on the board.
- Now ask what would happen to y if x doubles? (y halves)
- Next ask what would happen to x if y trebles? (x is divided by 3)
- Then ask what would happen if x is divided by 4? (y is multiplied by 4)
- Repeat with $xy = 36$ and other values.
- If time allows, discuss the connection with the inverse proportion questions.

Answers

Exercise 13C

1 **a** 6 hours **b** 4 hours **c** 6, 4, 5, 3, 2 **d** Product is always 600
e $xy = 600$ **f and g** Check pupils' graphs

2 **a** 100 **b** 200 **c** 500, 400, 200, 100, 50, 40
d Product is always the same, 1000 **e** $xy = 1000$
f and g Check pupils' graphs

3 **a** 5 hours **b** 1000 km/h **c** $xy = 4000$

4 **a** 24 **b** 24, 20, 12, 10 **c** Product is always 12 **d** $pn = 12$

5 **a** $bh = 200$ **b** 12.5, 16, 20, 25 **c** Check pupils' graphs
d About 13.3 **e** Formula gives $13\frac{1}{3}$

6 **a** £1500 **b** £1000 **c** 3000, 1500, 1000, 750, 600, 500
d Yes, the product is always the same (30000). **e** Check pupils' graphs
f About 38 **g** $nc = 30\,000$ The formula gives 37.5, which means 38 people

Activity: Different rectangles, same area

A Check pupils' drawings
B 12, 9.6, 8, 6, 4.8, 4
C Check pupils' graphs
D $xy = 48$
E About 6.9 cm

Lesson 13.4 Comparing direct proportion and inverse proportion

Learning objective

- To recognise direct and inverse proportion and work out missing values

Links to other subjects

- **Science** – to use wavelength and frequency graphs from experiments
- **Business studies** – to use financial graphs

Resources and homework

- Pupil Book 2.2, pages 239–241
- Homework Book 2, section 13.4
- Online homework 13.4, questions 1–10

Key words

- No new key words for this topic

Problem solving and reasoning help

- The **MR** questions in Exercise 13D of the Pupil Book test pupils' ability to determine whether questions involve direct or inverse proportion from both tables and graphs. Make sure that pupils have a good understanding of this by modelling several examples.

Common misconceptions and remediation

- Pupils will often use direct proportion rather than inverse proportion. Make sure that pupils are aware of the important words in a question, and the patterns of numbers in a table that indicate whether to use direct or inverse proportion. Also make sure that pupils write the variables the correct way around in their formula, as this is often a common error.

Probing questions

- How do you know whether given sets of numbers are in inverse proportion or direct proportion?
- What tips would you give someone to help them solve word problems involving inverse proportion or direct proportion?

Part 1

- Give pupils a ratio, such as 2 : 5, and ask them to find the missing value in the equivalent ratio … : 15 (6 : 15). Repeat with, for example, 10 : … (10 : 25), … : 35 (14 : 35).
- Pupils could use number fans or mini whiteboards for writing down their answers.
- Repeat with other ratios.

Part 2

- Discuss the following example, which is similar to Example 4 on page 239 of the Pupil Book), on direct and inverse proportion.
  Here are the value of two variables, x and y.

x	30	50
y	120	

a Find the missing value, when y is directly proportional to x.
Look at the values in the first column: $y \div x = 120 \div 30 = 4$. Therefore $y = 4x$ because they are in direct proportion. (When $x = 50$ then $y = 4 \times 50 = 200$.)

b Find the missing value, when y is inversely proportional to x. (Look at the values in the first column: $xy = 30 \times 120 = 3600$. Therefore $xy = 3600$ because they are in inverse proportion. When $xy = 3600$ and $x = 50$ then $50 \times y = 3600 \rightarrow y = 3600 \div 50 = 72$.)

- **Pupils can now do Exercise 13D from Pupil Book 2.2.**

Part 3

- Write the following problem on the board: $xy = 24$.
- Ask pupils what values x and y could take. You could write these on the board, as shown in the table.
- Then, as shown, draw the graph to illustrate inverse proportion on the board.

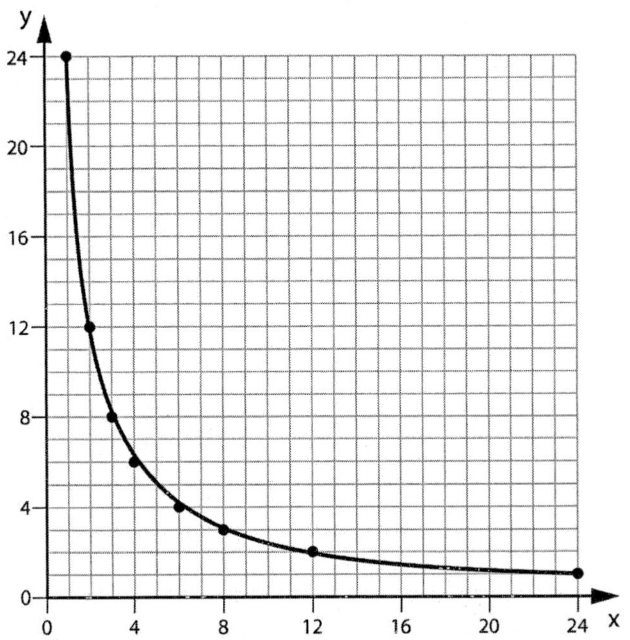

x	y
1	24
2	12
3	8
4	6
6	4
8	3
12	2
24	1

Answers

Exercise 13D

1 **a** Yes **b** $d = 80t$ **c** 680 m

2 **a** 4 hours **b** 2.5 hours **c** speed = 20 ÷ time so speed × time is always 20 **d** $tw = 20$

3 **a** Neither **b** Inverse $cd = 90$ **c** Direct $r = 13.5f$ **d** Neither

4 The line does not go through the origin

5 **a** (4, 5) and (7, 2) **b** 4 × 5 = 20 but 7 × 2 = 14; they are not the same

6 **a** $y = 0.25x$ or $y = \frac{1}{4}x$ **b** $xy = 400$

7 **a** Any two points, such as (10, 400) and (5, 200) **b** $y = 40x$

8 **a** Possible points are (12, 1), (1, 12), (3, 4) and (6, 2) **b** $xy = 12$

Reasoning: Looking for proportion

A 2 × (3 + 7) = 20

B Possible answers are 1 and 9, or 2 and 8, and so on

C Check pupils' graphs

D No, the line does not go to the origin

E No, it is not a curved line

Review questions (Pupil Book pages 242–243)

- The review questions will help to determine pupils' abilities with regard to the material within Chapter 12.
- These questions also draw on the maths covered in earlier chapters of the book to encourage pupils to make links between different topics.
- The answers are on the next page of this Teacher Book.

Challenge – Planning a trip (Pupil Book pages 244–245)

- For this challenge pupils are required to apply their understanding of proportion to a typical real-life context including speed, time and fuel consumption. The questions increase in complexity and pupils will need to use a range of graphical and algebraic skills to tackle the questions. Pupils also need to be able to interpret some quite complex language.
- Discuss question 1. Ask pupils:
  - Should the cost per person go down or up, the fewer people there are on the coach?
  - What calculation would you do to get £16 per person when 50 passengers are on the coach? (800 divided by 16)
  - How can you prove that the relationship between the number of passengers and the cost per person is inversely proportional? (Encourage pupils to represent this formally, that is: $y = \frac{800}{x}$).
- For part **c**, model the axes on the board for the class.
- For question 2, explain that 90 km/h means that the bus travels 90 km in one hour. Pupils should then be able to work out the values in the table.
- For part **d**, ask pupils:
  - How can you show that the values in the table are directly proportional? (Encourage pupils to represent this formally, that is: $y = 1.5x$).
- For part **e**, discuss the scale of the axes that pupils will use.
- For question 3, ask pupils to express the results from the table formally, that is: $y = \frac{240}{x}$. Then they can fill in the table.
- For part **c**, discuss the axes with pupils before they draw the graph.
- For question 4, ask pupils how they would decide whether the relationship is directly or inversely proportional? Encourage pupils to multiply and divide the values and to discuss their answers.
- Encourage pupils to model the relationships formally.

Answers to Review questions

1 34 and 85

2 8 cm, 32 cm, 80 cm, 2.4 m, 6.4 m

3 540

4 **a i** £1.40 **ii** £6 **iii** £73 **b** $c = 0.2n$

5 **a** 28, 70, 7 **b** $y = 3.5x$ **c** Check pupils' graphs
d Straight line through the origin

6 9 and 3

7 **a** 400 seconds **b** 67 seconds

8 **a** length × width is always 1200 **b** 30 cm
c $xy = 1200$ **d** Check pupils' graphs

9 **a** 100, 75, 60 **b** $ln = 180$

10 **a i** (5, 1) **ii** (10, 0.5) **iii** (2, 2.5) **iv** (1, 5)
b $xy = 5$ **c** 0.2

11 **a** 32 **b** 2

Answers to Challenge – Planning a trip

1 **a** 20, 25, 32, 40
b $xy = 800$
c

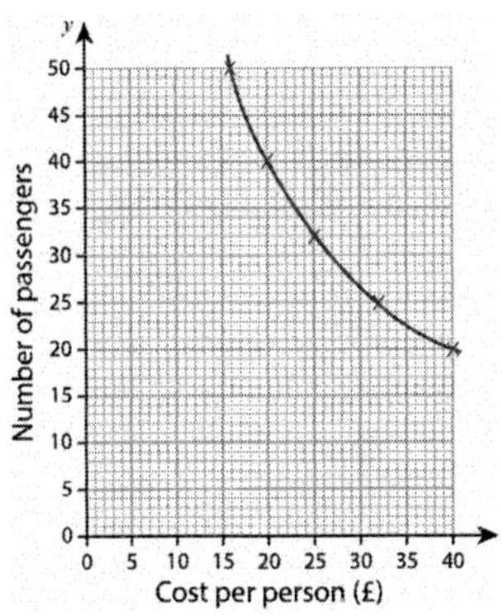

2 **a** 45 km
b 15 km
c 15, 30, 45, 60, 75, 90
d yes, $d = 1.5t$
e

3 **a** 3.2, 3, 2.4

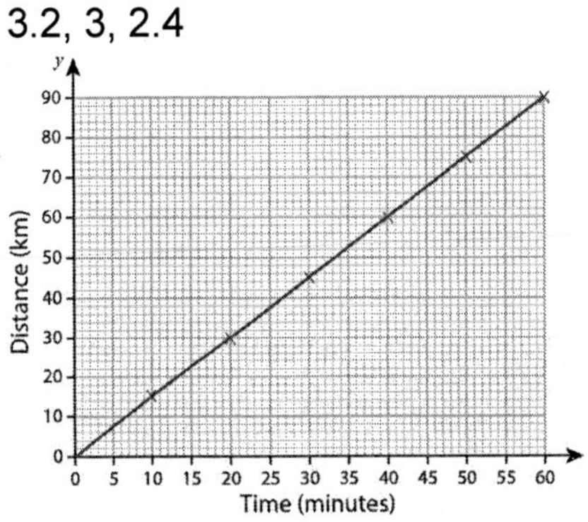

b $xy = 240$
c

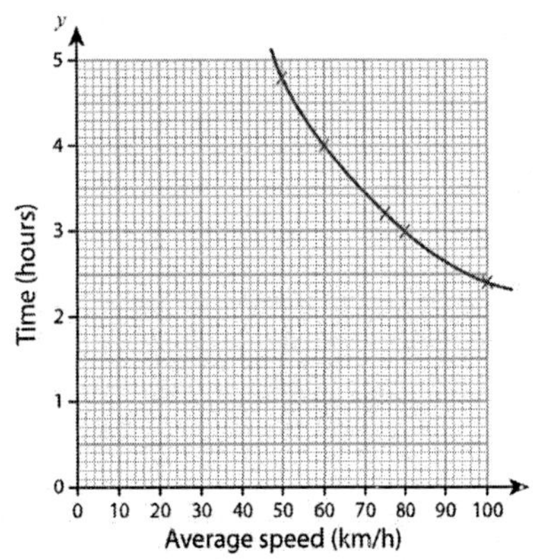

4 **a** i direct proportion
b $d = 2.5f$
c iii neither. Check pupils' reasons

14 Circles

Learning objectives

- The names of the parts of a circle
- How to calculate the circumference of a circle
- How to calculate the area of a circle

Prior knowledge

- The words 'radius' and 'diameter'
- How to use compasses to draw a circle
- What a regular polygon is

Context

- Tell pupils that the circle is probably the most important shape in the universe. It is also the most mysterious. We use a fascinating number that pupils may have heard of, called pi, written as π, which is used to calculate the circumference (perimeter) of a circle. But π cannot be written exactly as a number and its decimal places never end. Pupils could prepare for this chapter by doing their own research on π. Encourage pupils to present their findings to the class.

Discussion points

- Ask pupils to name the radius, diameter and circumference of a circle.
- What is the relationship between the diameter and the radius of circle?

Associated Collins ICT resources

- Chapter 14 interactive activities on Collins Connect online platform
- *Clock* and *Trapeze artists* Wonder of Maths on Collins Connect online platform

Curriculum references

Develop fluency

- Use language and properties precisely to analyse numbers, algebraic expressions, 2D and 3D shapes, probability and statistics

Solve problems

- Develop their mathematical knowledge, in part through solving problems and evaluating the outcomes, including multi-step problems

Geometry and measure

- Calculate and solve problems involving: perimeters of 2D shapes (including circles), areas of circles and composite shapes
- Derive and illustrate properties of triangles, quadrilaterals, circles, and other plane figures [for example, equal lengths and angles] using appropriate language and technologies

Fast-track for classes following a 2-year scheme of work

- Pupils may be familiar with the content of Lesson 14.1 and Lesson 14.2. Check pupils' understanding with some examples. If pupils are confident and fluent move straight on to Lesson 14.3 and Lesson 14.4.

Lesson 14.1 The circle and its parts

Learning objective

- To know the definition of a circle and the names of its parts

Resources and homework

- Pupil Book 2.2, pages 247–249
- Homework Book 2, section 14.1

Links to other subjects

- **Design and technology** – to use the parts of a circle to describe the characteristic of a design involving circles
- **Physical Education** – to use the language of circles to describe the features of a track or the patterns in a dance routine

Key words

- arc
- centre
- chord
- circumference
- diameter
- radius
- sector
- segment
- semicircle
- tangent

Problem solving and reasoning help

- **MR** question 5 in Exercise 14A of the Pupil Book and the activity at the end of the exercise give pupils the opportunity to use the language and characteristics of a circle that they have already learned. Pupils will do more complex construction and problem-solving activities. Use guided group work to support **less able** pupils.

Common misconceptions and remediation

- Pupils often confuse radius and diameter. Give them plenty of opportunity to use both. Collective memory activities would be useful in helping pupils to remember which is which.

Probing questions

- Visualise and identify a 3D shape whose front and side elevations are both rectangles and the plan is a circle.

Part 1

- Draw a circle on the board, marking the centre clearly. Ask pupils to write a short definition of a circle. Share pupils' examples, then compare it to this example: *A two-dimensional shape made by drawing a curve that is always the same distance from a centre.*
- Now compare it to the definition on page 247 of the Pupil Book: A circle is a set of points that are all the same distance from a fixed point, called the centre.

Part 2

- In this lesson, pupils will need to learn and recall the names of parts of a circle, so that they will know what is being asked of them in questions such as those later in this chapter.
- Ask pupils to work in groups of five or six for a reading image activity.
- Copy the examples below on an A3 sheet, laminate it and place it at the back of the class. Remove the names and definitions, photocopy it and hand out a sheet to each group.
- Ask one group member to go to the back of the class and study the completed example for about one minute. Then that person must return to the group and provide them with clear descriptions of each circle part so that group members can reproduce the completed example. Repeat until each group member has had a turn. Groups can then swap examples and mark each other's work formatively using a projection of the completed version.

- Support or challenge pupils with: timing, repetition and grouping. Support **less able** pupils by using probing questions or partly completed eamples.

Circumference: the distance around the circle; a special name for the perimeter of a circle

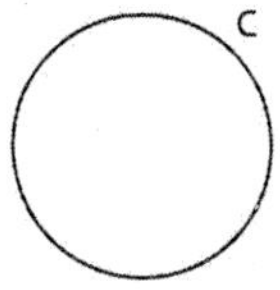

circumference

Chord: a line that cuts the circle into two parts
Tangent: a line that touches the circle at a single point on the circumference

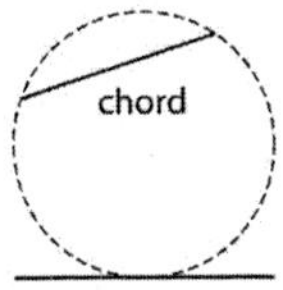

tangent

Arc: a part of the circumference

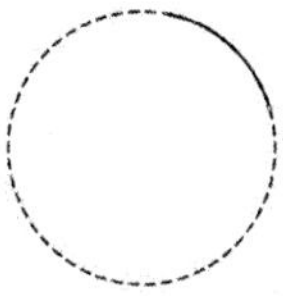
arc

Segment: an area of the circle enclosed by a chord and an arc

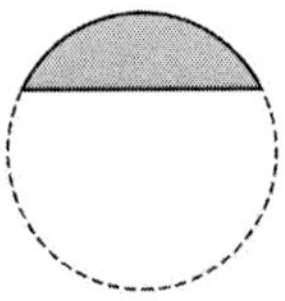
segment

Radius (*r*): the distance from the centre to the circumference; the plural is 'radii'

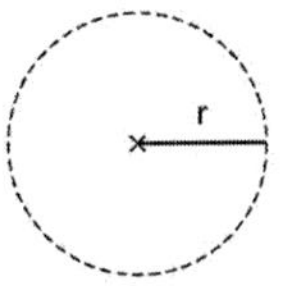

radius

Sector: an area of the circle enclosed by two radii and an arc

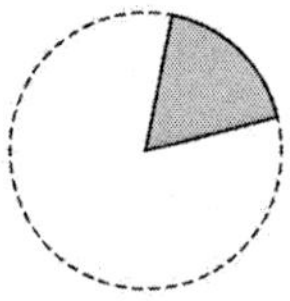
sector

Diameter (*d*): the distance from side to side passing through the centre; the greatest width of a circle; *d* is twice the *r*, $d = 2r$

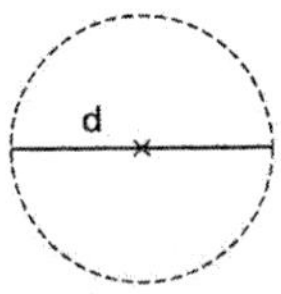

diameter

Semicircle: half a circle; the area of the circle enclosed by a diameter and an arc

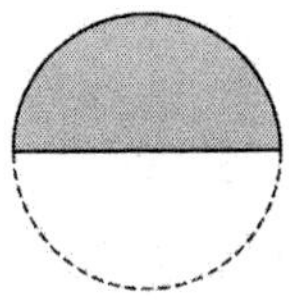
semicircle

- **Pupils can now do Exercise 14A from Pupil Book 2.2.**

Part 3

- Challenge: How quickly can you complete a blank copy of the diagram used in Part 2?

Answers

Exercise 14A

1 **a** 1.5 cm, 3 cm **b** 2 cm, 4 cm **c** 3 cm, 6 cm

2-4 Check pupils' drawings.

5 **a** 360° ÷ 6 = 60° **b i** 360° ÷ 5 = 72° **ii** 360° ÷ 8 = 45° **iii** 360° ÷ 10 = 36°

Activity: Finding the centre of a circle

Check pupils' drawings.

Lesson 14.2 Circumference of a circle

Learning objective

- To work out the relationship between the circumference and diameter of a circle

Links to other subjects

- **Design and technology** – to use the definition of the circumference of a circle to describe the characteristic of a design involving circles
- **Physical Education** – to use the language of circles to describe the features of a track
- **Science** – to use the definition of circumference to approximate the solution of distance–time problems involving circles in real life

Resources and homework

- Pupil Book 2.2, pages 250–251
- Homework Book 2, section 14.1

Key words

- No new key words for this topic

Problem solving and reasoning help

- The activity in Exercise 14B in the Pupil Book forms most of Part 2 of this lesson. It explains the common approach to discovering the relationship between the circumference and diameter of a circle. The activity at the end of Exercise 14B gives pupils the opportunity to extend their understanding of circles to help them solve problems involving 3D shapes.

Common misconceptions and remediation

- Pupils often do not make the link between the work they have done previously on perimeter and area and the work on the circumference and area of a circle.

Probing questions

- A circle has a diameter of 10 m. What is its circumference?
- A circle has a circumference of 140 cm. What are its diameter and its radius?

Part 1

- Start by revising the perimeters and areas, using familiar shapes such as rectangles.
- Make sure that each pair of pupils has an individual whiteboard or a sheet of A4 paper.
- Ask pairs to draw a rectangle that has a perimeter of 10 cm and an area of 6 cm^2. Once done, ask each pair to hold it up. The first pair to show the correct answer gains a point.
- Repeat the activity using different examples to suit the ability of the class.

Part 2

- Explain to the class that they are going to do a practical activity to see if there is a relation between the diameter and the circumference of a circle.
- It might be a good idea for pupils to work in pairs. Each pair will require a pair of compasses, a 30-cm ruler and a piece of string about 40 cm in length.
- **Pupils can now do Exercise 14B from Pupil Book 2.2.**

Part 3

- Summarise what the class have discovered during the lesson.
- Explain that a simple relation between the circumference C and the diameter d of a circle is $C = 3d$.
- Explain that in the next lesson they will be given a more accurate equation.

Answers

Exercise 14B

Circumference divided by diameter is slightly larger than 3. A simple relationship is $C = 3d$.

Activity: Making nets for cones

As the size of the removed sector increases, so does the height of the cone.

Lesson 14.3 Formula for the circumference of a circle

Learning objective

- To calculate the circumference of a circle

Resources and homework

- Pupil Book 2.2, pages 251–254
- Homework Book 2, section 14.2
- Online homework 14.3, questions 1–10

Links to other subjects

- **Design and technology** – to use the formula of the circumference of a circle to complete designs incorporating circles
- **Physical Education** – to use the language of circles to describe the features of a track
- **Science** – to use the definition of circumference to approximate the solution of distance–time problems involving circles in real life

Key word

- π

Problem solving and reasoning help

- **PS** question 7 of Exercise 14C in the Pupil Book requires pupils to apply what they have learnt to a multi-step problem. **MR** question 8 requires pupils to apply their understanding to a familiar real life problem. The activity at the end of the exercise gives pupils the opportunity to do some independent research about π. Pupils may need guidance on how to conduct their own independent research, including comparing a number of different sources instead of relying on the first information they find.

Common misconceptions and remediation

- Pupils often confuse the definitions of the parts of circles. Use activities like the one in Lesson 14.1 to help them overcome this. Pupils also confuse the formulae, particularly in terms of the use of radius and diameter. This is often because they cannot visualise what they are doing. Activities like those in the previous lesson will help pupils to overcome this.

Probing questions

- A touring cycle has wheels of diameter 75 cm. How many rotations does each wheel make for every 10 km travelled?
- Give pupils some work with mistakes. Ask them to identify and correct the mistakes.

Part 1

- Ask pupils to remember the relationship they found in the last lesson: $C = 3d$.
- Using this starting point, ask pupils if they can write a similar formula for the circumference of a circle in terms of its radius? What about the formula for half a circle (a semicircle)?
- *Challenge*: Remind pupils about the definition of a sector and of an arc (refer to Lesson 14.1 or page 247 of the Pupil Book). If the angle at the centre of the circle is 60° and they know the circumference could they write a similar relationship for the length of the arc?

Part 2

- Draw a circle on the board or OHP. Recap: *radius*, *diameter* and *circumference*.
- Explain that it is possible to find the circumference and area of the circle by using formulae, and that in Year 9 they will be shown how to find the formulae.
- Tell pupils that they will deal with the formula for the area of the circle in the next lesson. The formula for the circumference of a circle is given by $C = \pi d$, where C is the circumference and d is the diameter of the circle. The Greek letter π (pi) is the symbol that is used for circles. The value of π cannot be written down exactly, so approximate values are used. It is usual to take π as 3.14 to two decimal places, or to use the π key on a calculator.
- Draw two circles and show the class how to calculate the circumference and area, given either the diameter or the radius, for example: Given the diameter; Given the radius.

A

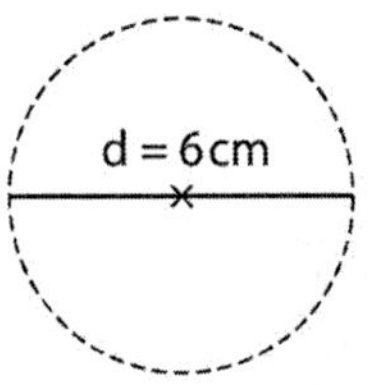

d = 6cm, so r = 3cm

$C = \pi d$

$= \pi \times 6$

= 18.8 cm (to 1 dp)

B

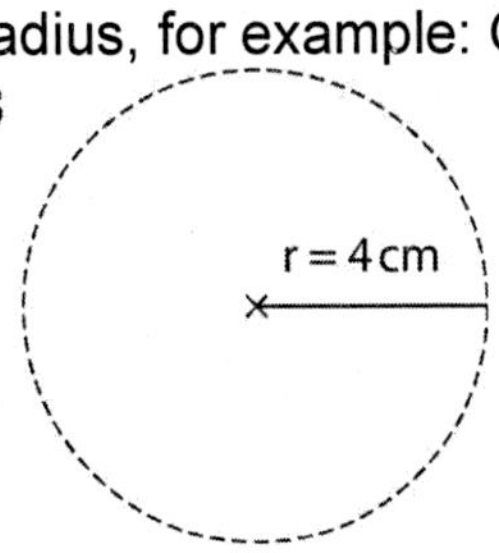

r = 4cm, so d = 8cm

$C = \pi d$

$= \pi \times 8$

= 25.1 cm (to 1 dp)

- **Pupils can now do Exercise 14B from Pupil Book 2.2.**

Part 3

- Ask pupils to work in small groups on the following problem:
  Sarah and Jane each have a circular rug in their bedrooms.
  Sarah's rug has a diameter of 3 m. Jane's rug has a diameter of 1.5 m.
  Sarah says that the circumference of her rug is twice the circumference of Jane's.
  Is she correct? Explain your answer.
- Encourage pupils to focus on explaining their answer. Then identify one group or more to present their answer to the class. The rest of the class could give formative feedback based on a set of agreed criteria. For example: accuracy of solution, quality of explanation, use of diagrams and formulae to support their argument.

Answers

Exercise 14C

(note: answers could be slightly different if value of π is taken as 3.14)

1 **a** 22.0 cm **b** 34.6 mm **c** 66.0 mm **d** 15.1 m **e** 8.8 cm

2 82 mm

3 377 m

4 400 m

5 942 million km

6 12.9 cm

7 47.1 cm

8 32 m

Activity: A mnemonic for π

Count the number of letters in each word:

3 1 4 1 5 9 2 7 which leads to 3.1415927

Lesson 14.4 Formula for the area of a circle

Learning objective

- To calculate the area of a circle

Resources and homework

- Pupil Book 2.2, pages 254–257
- Homework Book 2, section 14.3
- Online homework 14.4, questions 1–10

Links to other subjects

- **Design and technology** – to use the formula of the area of a circle to calculate the raw material needed to produce an article for a design template
- **Physical Education** – to use the formula for the area of a circle to calculate the dimensions of different sprint tracks

Key words

- No new key words for this topic

Problem solving and reasoning help

- The mathematical reasoning and problem solving questions in this lesson require pupils to apply their learning to more complex problem solving questions. If necessary model an example to help them understand how to identify the important information and the mathematics needed to answer complex multi-step problems.

Common misconceptions and remediation

- Pupils often confuse the definitions of the parts of circles use activities like the one in lesson 1 to help overcome this.
- They also confuse the formula particularly in terms of the use of radius and diameter. This is often because they cannot visualise what they are doing. Activities those in the last lesson will help them overcome this.

Probing questions

- A circle has a radius of 20 cm. What is its area?
- A door is a rectangle shape with a semicircular arch on top. The rectangular part is 2 m high and the door is 90 cm wide. What is the area of the door?
- What information do you need to be able to find the circumference and area of a circle?
- How would you go about finding the area of a circle if you know the circumference?

Part 1

- Ask pupils to explain what is the same, and what is different about the formulae for the circumference and for the area of a circle.
- Can they write a formula for the area of a circular in terms of the circumference instead of the radius and diameter?

Part 2

- Remind pupils of the example they used in Lesson 14.3 to work out the circumference of a circle, using $C = \pi d$, where C is the circumference and d is the diameter of the circle.
- Say that the formula for the area of a circle is given by $A = \pi r^2$, where A is the area and r is the radius of the circle. Explain that πr^2 is the same as $\pi \times r \times r$ and they must be careful when using the formula on a calculator.

- Draw two circles and show the class how to calculate the circumference and area, given either the diameter or the radius, for example:
  Given the diameter; Given the radius.

A B

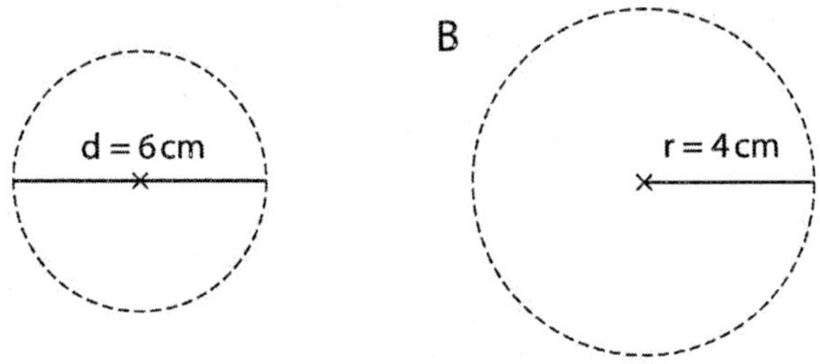

$A = \pi r^2$
$= \pi \times 3 \times 3$
$= \pi \times 9$
$= 28.3$ cm^2 (to 1 dp)

$A = \pi r^2$
$= \pi \times 4 \times 4$
$= \pi \times 16$
$= 50.3$ cm^2 (to 1 dp)

Part 3

- Ask pupils to write down the formulae to find the circumference and area of a circle. Emphasise the need to learn these formulae, as they are not given in examinations.
- *Challenge*: The diagram shows a triangle and three circles whose centres are at the vertices of the triangle. The area of the triangle is 100 cm^2; each circle has a radius of 3 cm. What is the area, in cm^2, of the shaded area?

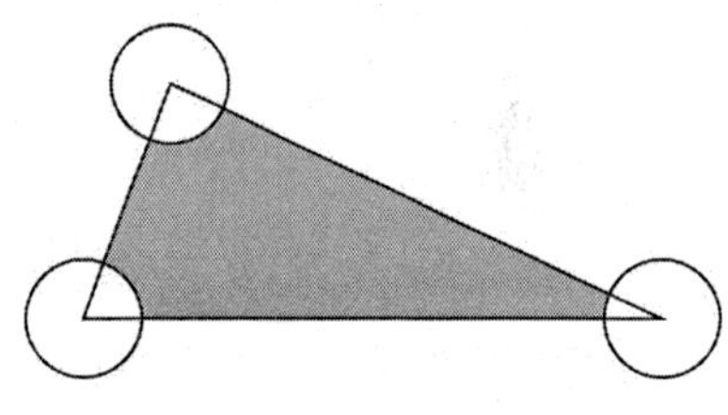

- Use these hints as appropriate:
  - The three angles of the triangle add to 180°.
  - A sector of a circle is the area of a circle enclosed by two radii and an arc.
  - The combined area of the three sectors of the circles that are inside the triangle must add up to half a circle. The area of the half circle must be $\frac{1}{2} \times 32\ \pi$.

Answers

Exercise 14D

(note: answers could be slightly different if value of π is taken as 3.14)

1 **a** 3.1 cm^2 **b** 153.9 mm^2 **c** 3.5 m^2 **d** 38.5 cm^2 **e** 95.0 m^2
2 346 cm^2
3 20 mm, 314 mm^2
4 He has worked out the circumference of the circle and the units are wrong.
Area = $\pi \times r^2 = \pi \times 16 = 16\pi$ cm^2
5 8163 m^2
6 169π cm^2
7 47.5 cm^2
8 25π cm^2 and 100π cm^2 No, the area is 4 times larger.

Problem solving: Circle problems

A **a** 75.4 cm^2 **b** 60.7 cm^2 **c** 63.3 cm^2
B 8.0 m
C 17.9 cm^2

Review questions (Pupil Book pages 258–259)

- The review questions will help to determine pupils' abilities with regard to the material within Chapter 14.
- These questions also draw on the maths covered in earlier chapters of the book to encourage pupils to make links between different topics.
- The answers are on the next page of this Teacher Book.

Financial skills – Athletic stadium (Pupil Book pages 260–261)

- This activity is designed to give pupils the opportunity to apply their knowledge to a multi-step real-life problem. The context is familiar, but the activity is presented in a slightly more complex way than pupils may be used to.
- All the information required to answer the questions is in the text, but pupils will need to read and then think carefully about how they access the information. Remind them to highlight the key information they will need. Tell pupils that they will need to combine their skills not only from this chapter, but also from different areas of mathematics, for example, number.
- Before the class starts working on questions 1 to 6 in the Pupil Book, ask them some questions as a warm-up. Here are some examples:
    - Can you think of everyday examples of when you might need to use the formulae for the circumference and/or area of a circle?
    - Can you think of any sports that use circles in the design of their tracks and/or pitches?
- Pupils can now work on the questions individually or in groups.
- Encourage pupils to develop this topic further by using the internet to research different sports. Groups or individuals could present their findings to the class.

Answers to Review questions

(note: answers could be slightly different if value of π is taken as 3.14)

1	Check pupils' drawings			
2	**a i** 56.5 mm	**ii** 254.5 mm^2	**b i** 22.0 cm	**ii** 38.5 cm^2
	c i 15.1 m	**ii** 18.1 m^2		
3	**a** 44 cm	**b** 154 cm^2		
4	**a** 157.1 cm	**b** 134 m		
5	**a** 9.4 cm	**b** 1060		
6	**a** 66 cm	**b** 207 cm	**c** 4830	
7	30.5 cm^2			
8	**a** 200	**b** 25 cm^2	**c** 12.6 cm^2	**d** 49.6%

Answers to Financial skills – Athletic stadium

(note: answers could be slightly different if value of π is taken as 3.14)

1	**a** 6.72 m	**b** 3.60 m^2	**c** £115.20	
2	**a** 353.4 m^2	**b** £20 320		
3	**a** 399.34 m (or 400 m)		**b** 2000 litres	**c** £2635
4	**a** 10 400m^2	**b** £37 440		
5	**a** 9.6 m^3	**b** 16 tonnes		
6	**a** 0.9 m^3	**b** £43.20		

15 Equations and formulae

Learning objectives

- How to solve equations involving brackets and fractions
- How to solve equations where the variable occurs more than once
- How to rearrange formulae

Prior knowledge

- How to solve simple equations
- How to write simple formulae

Context

- This chapter starts by introducing brackets to the simple equations that pupils have solved previously. Pupils are then shown how to solve equations with unknowns on both sides that also include brackets and fractions. Finally changing the subject of a formula is tackled.

Discussion points

- How would you construct an equation from the information given in a problem?
- How would you check if the equation will work?
- What are the steps taken when solving an equation?

Associated Collins ICT resources

- Chapter 15 interactive activities on Collins Connect online platform
- *Solving equations with the unknown on both sides* video on Collins Connect online platform
- *Stadium* Wonder of Maths on Collins Connect online platform

Curriculum references

Develop fluency

- Substitute values in expressions, rearrange and simplify expressions, and solve equations

Solve problems

- Develop their mathematical knowledge, in part through solving problems and evaluating the outcomes, including multi-step problems

Algebra

- Understand and use standard mathematical formulae; rearrange formulae to change the subject
- Use algebraic methods to solve linear equations in one variable (including all forms that require rearrangement

Fast-track for classes following a 2-year scheme of work

- Much of this chapter will be new material. However, pupils who are familiar with multiplying out brackets and solving simple equations will be able to complete Exercise 15A in the Pupil Book quickly before moving on to Exercise 15B. Or, you could suggest that these pupils leave out Exercise 15A altogether and start with Exercise 15B.

Lesson 15.1 Equations with brackets

Learning objective

- To solve equations involving brackets

Resources and homework

- Pupil Book 2.2, pages 263–264
- Homework Book 2, section 15.1
- Online homework 15.1, questions 1–10

Links to other subjects

- **Science** – to solve equations in physics
- **Business studies** – to solve financial calculations

Key word

- inverse operation

Problem solving and reasoning help

- The challenge activity at the end of Exercise 15A in the Pupil Book enables pupils to demonstrate their learning. Encourage **less able** pupils to follow the methods shown to help them to try and work out the answers to each question. You could ask **more able** pupils to design their own questions.

Common misconceptions and remediation

- A common problem that pupils have when expanding a bracket is to multiply the first term by the number outside the bracket and write down only the second term. Another problem is that sometimes pupils add instead of multiplying. Use the activity in Part 1 to ensure that pupils can expand brackets correctly before tackling Exercise 16A in the Pupil Book.

Probing questions

- Show pupils a list of different linear equations with brackets and ask:
  - Which equations can be solved easily?
  - Which equations are more difficult to solve and why?
  - What strategies should you use with the more difficult equations? Explain why.

Part 1

- Write these on the board and ask pupils to expand the brackets:
  - $3(x + 4)$
  - $2(x - 5)$
  - $3(2x + 4)$
  - $\frac{1}{3}(6x - 3)$
- Now show pupils examples of errors when multiplying out brackets with errors. Ask pupils to explain what the errors are in these equations:

  $3(x +2) = 3x + 2$

  $2(x - 3) = 2x - 5$

  $-2(6 - x) = -12 - 2x$

Part 2

- Write the equation $3(3x + 1) = 21$ on the board. Ask pupils how they would start to solve it.
- If pupils suggest multiplying out the bracket first, then put this and the rest of the solution on the board ($x = 2$), asking pupils for the next step at each stage.
- Now ask if there is a different way to start to solve the equation. The alternative you are looking for is to start by dividing both sides by 3 to give: $3x + 1 = 7$.

- Work through the solution to obtain the same answer as with the previous method.
- **Pupils can now do Exercise 15A from the Pupil Book.**

Part 3

- Ask if anyone can remember the name of the type of equation they have been looking at today. (Linear equations)
- Explain that this sort of equation crops up in many different areas of mathematics, science and technology. Being able to solve equations like this (and more complicated equations) is essential to being able to solve problems in these subjects.
- Try to get across the idea that solving equations is just like solving a puzzle. The difference is that when solving equations, we try to use a logical sequence to help us solve the puzzle, which is guaranteed to give the correct answer every time if we do it correctly.

Answers

Exercise 15A

1	**a** 33	**b** 48	**c** 6	**d** 4				
2	**a** 13	**b** 9	**c** $5\frac{3}{4}$	**d** $5\frac{2}{3}$	**e** $7\frac{4}{7}$	**f** $3\frac{1}{3}$	**g** $2\frac{3}{5}$	**h** $4\frac{5}{8}$
3	**a** 11	**b** 3	**c** 18	**d** 7	**e** 15	**f** 7	**g** 17	**h** 44
4	**a** 48	**b** 36	**c** 45	**d** 32				
5	**a** 36	**b** 18	**c** 30	**d** 58	**e** 6	**f** 13	**g** 27	**h** 22
6	Both give $x = 8\frac{2}{3}$							
7	**a** $6\frac{3}{4}$	**b** $5\frac{1}{4}$	**c** $6\frac{1}{2}$	**d** $1\frac{1}{2}$	**e** $2\frac{1}{4}$	**f** 3	**g** $6\frac{3}{4}$	**h** $6\frac{2}{5}$
8	**a** $3x + 12\frac{1}{2} = 37\frac{1}{2}$		**b** $x = 8\frac{1}{3}$					
9	**a** 3.6	**b** 7.1	**c** 12.4	**d** 2.8	**e** 16.1	**f** 3.8	**g** 0.4	**h** 62.5

Challenge: Odd one out

The odd one out is $5x - 30 = 75$. This has the answer 21, the rest are all 19.

Lesson 15.2 Equations with the variable on both sides

Learning objective

- To solve equations with the variable on both sides

Resources and homework

- Pupil Book 2.2, pages 265–267
- Homework Book 2, section 15.2
- Online homework 15.2, questions 1–10

Links to other subjects

- **Science** – to solve equations in physics
- **Business studies** – to solve financial calculations

Key words

- No new key words for this topic

Problem solving and reasoning help

- **PS** question 7 in Exercise 15B of the Pupil Book is a word question that requires pupils to formulate and write down an equation before solving it. **Less able** pupils may struggle with this type of question so you may need to model some examples for them, picking out the key words from the question before they attempt to solve it.

Common misconceptions and remediation

- Pupils will sometimes get confused with adding or subtracting from each side when dealing with equations with unknowns on both sides. To help **less able** pupils, explain that in order to remove a negative value you add to both sides; to remove a positive value you subtract from both sides. Pupils who struggle may benefit from following these steps:
    1. Remove the smallest x value by adding or subtracting from both sides (explain, for example, that $-5x$ is smaller than $2x$).
    2. Remove the number on the same side as the remaining x value by adding or subtracting from both sides.
    3. Finally, divide both sides by the number in front of the x to obtain your answer.

Probing questions

- Show pupils a list of different linear equations with unknowns on both sides and ask:
    - Which can be solved easily?
    - Which are more difficult to solve and why?
    - What strategies should you use with the more difficult ones and why would these strategies make solving the equations easier?

Part 1

- Write the following on the board: +, –, ×, ÷, √, 5, 55, 555, 5555, ( ).
- Ask pupils to make as many different integers as possible. Say that they must use the digit 5 exactly four times in each calculation. Write these examples on the board:
- $5 + 5 + 5 + 5 = 20$; $(55 + 5) \div 5 = 12$; $55 \div 55 = 1$; $55 + (5 \div 5) = 56$; $\sqrt{55} \times \sqrt{55} = 55$.
- Once everyone has grasped the idea, set the task of trying to make as many of the integers from 1 to 10 as possible using this method. Possible solutions are:

1 $55 \div 55$

2 $(5 \div 5) + (5 \div 5)$

3 $5 - \frac{(5 + 5)}{5}$

4 $\sqrt{5} \times \sqrt{5} - \frac{5}{5}$

5 $\sqrt{5} \times \sqrt{5} \times \frac{5}{5}$

6 $\sqrt{5} \times \sqrt{5} + \frac{5}{5}$

7 $5 + \frac{(5 + 5)}{5}$

8 $\frac{(5 + 5 + 5)}{5} + 5$ (8 cannot actually be generated without using five 5s)

9 $5 + 5 - \frac{5}{5}$

10 $(5 + 5) \times \frac{5}{5}$

- Encourage pupils who complete this early to try and generate the integers up to 20.

Part 2

- Write the equation $6x = 20 + 2x$ on the board. Ask pupils what is different about this equation to those they have met before. Someone may notice that there is an unknown on both sides.
- Say that we need to have an unknown on one side only, so that it ends up as a positive term rather than as a negative term. To do this here we need to subtract $2x$ from both sides:
  $6x - 2x = 20 + 2x - 2x$
  $4x = 20$
- Now we have a familiar type of equation, which we can solve to get: $x = 5$.
- Write another equation on the board: $5x - 4 = 12 - 3x$. Ask pupils how to solve this equation.
- This time we need to add $3x$ to both sides: $5x - 4 + 3x = 12 - 3x + 3x$
  $x - 4 = 12$

  Continue the solution to give: $x = 2$.
- **Pupils can now do Exercise 15B from Pupil Book 2.2.**

Part 3

- Invite pupils to the board to solve examples of the types of equation covered in this lesson:
  $3x + 4 = x + 16$; $x - 5 = 8x + 2$; $6x + 6 = 4x + 12$
- Extend to more complex examples; link Lesson 15.1 and Lesson 15.3 for **more able** pupils:
  $4(3 + x) = 3(x - 4)$; $3x + 3 - 2x = 6(x + 3)$

Answers

Exercise 15B

1	**a** 15	**b** 15	**c** 12	**d** 11		
2	**a** 9	**b** 6	**c** 4.5	**d** 18		
3	**a** 20	**b** 7	**c** 11	**d** 5.5		
4	**a** 8	**b** 21	**c** 6	**d** 8		
5	**a** 18	**b** 10.5	**c** $3\frac{1}{3}$	**d** 34	**e** 26	**f** $\frac{3}{4}$
	g 2	**h** 8	**i** 7	**j** 13 **k** 5	**l** 20	
6	**a** $2x + 35 = 3x + 12$			**b** $x = 23$	**c** 81	
7	**a** $n + 75$	**b** $n + 75 = 4n$		**c** $n = 25$; Ann has £25, Carrie has £100		
8	**a** 14	**b** 9	**c** 10	**d** $5\frac{1}{2}$		
9	**a** 12	**b** 8	**c** 6	**d** 4		
10	**a** 15	**b** 9	**c** 4	**d** 12		
	e 4	**f** 24	**g** 2	**h** 5		
11	**a** $4\frac{1}{2}$	**b** $2\frac{1}{4}$	**c** $6\frac{1}{2}$	**d** $3\frac{1}{4}$		

Challenge: Muddying the waters

A Missing numbers are 21, 26, 17, 32

B Missing expressions are $17 + 5x$, $17 + 9x$, $17 + 14x$ and $15x + 4$

Lesson 15.3 More complex equations

Learning objectives

- To solve equations with fractional coefficients
- To solve equations with brackets and fractions

Resources and homework

- Pupil Book 2.2, pages 267–270
- Homework Book 2, section 15.3
- Online homework 15.3, questions 1–10

Links to other subjects

- **Business studies** – to solve problems involving money

Key words

- No new key words for this topic

Problem solving and reasoning help

- The **PS** questions in Exercise 15C of the Pupil Book require pupils to formulate and solve equations from either 2D shapes with unknown side lengths or from word questions. Ensure that pupils can work out the area and perimeter of triangles, squares and rectangles before they attempt questions 7 and 8. **Less able** pupils may find setting up of equations difficult. Therefore it is may be helpful to use values for the unknowns showing the structure of each equation. **Less able** pupils may also struggle with word questions such as question 10. If so, model some examples, picking out the key words from the question before pupils attempt it.

Common misconceptions and remediation

- A common problem that pupils have when expanding a bracket is to multiply the first term by the number outside the bracket and write down only the second term. Another problem is that sometimes pupils add instead of multiplying. When solving equations that contain fractions pupils will sometimes remove the fraction by multiplying but leaving the denominator next to the bracket, for example:

  $\frac{1}{4}(x + 2) = 5$ is sometimes *incorrectly* written as $4(x + 2) = 20$

Probing questions

- Show pupils a list of different linear equations involving brackets and fractions and ask:
  - Which equations can be solved easily?
  - Which equations are more difficult to solve and why?
  - What strategies should you use with the more difficult equations and why would these strategies make solving the equations easier?

Part 1

- Ask for some equivalent fractions to $\frac{1}{2}$. After a few correct suggestions such as: $\frac{2}{4}, \frac{5}{15}$ … , ask for an equivalent fraction to $\frac{1}{2}$ that uses 34. The two possible answers are $\frac{34}{68}$ and $\frac{17}{34}$.
- Now ask for two fractions that are equivalent to $\frac{1}{3}$, which use the number 12 ($\frac{12}{36}$ and $\frac{4}{12}$).
- Repeat this with the following examples:
  - Equivalent to $\frac{1}{5}$ using the number 45. ($\frac{45}{225}$ and $\frac{9}{45}$)
  - Equivalent to $\frac{2}{3}$ using the number 18. ($\frac{18}{27}$ and $\frac{12}{18}$)
  - Equivalent to $\frac{1}{4}$ using the number 28. ($\frac{28}{112}$ and $\frac{7}{28}$)
  - Equivalent to $\frac{3}{4}$ using the number 36. ($\frac{36}{48}$ and $\frac{27}{36}$)

Part 2

- Write the equation on the board: $5(x - 2) = 3(x + 4)$. Ask pupils how to solve this equation.
- We need to multiply out the brackets first to give: $5x - 10 = 3x + 12$
- Next we need to subtract $3x$ from both sides:
  $5x - 10 - 3x = 3x + 12 - 3x$
  $2x - 10 = 12$
- Continue the solution to give $x = 11$.
- Write the equation $\frac{1}{3}x = 5$ on the board, and ask the class how to solve it.
- Pupils should respond with: 'multiply both sides by 3', which gives: $x = 15$.
- Then write down the equation $\frac{1}{3}(4x + 5) = 7$ and ask 'How do we solve this?'
- Again, it is a matter of simplifying, step-by-step, the side that has the variable x.
- Start by multiplying both sides by 3 to give $4x + 5 = 21$. Then subtract 5 from both sides to give $4x = 16$. Finally, divide both sides by 4 to give $x = 4$.
- **Pupils can now do Exercise 15C from Pupil Book 2.2.**

Part 2

- Invite pupils to the board to solve examples of equations covered in this lesson:

a $3(x + 5) = 21 + x$
b $4(2x + 3) = 26 + x$
c $4(3x - 1) = 6 + 7x$
d $(x - 6) = 2(1 + x)$
e $8x + 9 = 3(5 + 2x)$
f $7x - 2 = 2(1 + 3x)$

Answers

Exercise 15C

1	**a** 9	**b** 8	**c** 18	**d** 2				
2	**a** 24	**b** 9	**c** 8	**d** 21	**e** 22	**f** 8	**g** 7	**h** 8
3	**a** Top row 14, 16, 18, 20		bottom row 12, 15, 18, 21			**b** $x = 8$	**c** Pupils check own	
4	**a** Top row 16, 20, 24, 28		bottom row 30, 27, 24, 21			**b** $x = 4$	**c** Pupils check own	
5	**a** 24	**b** 21	**c** 14	**d** 8				
	e 16	**f** 6	**g** 5	**h** 0	**i** 20.5			
6	**a** $5(x - 4) = 3(x + 2)$		**b** $x = 13$	**c** 45 cm				
7	**a** $3(t + 6)$	**b** $4t$	**c** $3(t + 6) = 4t$		**d** $t = 18$			
	e Triangle 24 units, square 18 units							
8	**a** $2(x + 5)$	**b** $4(x - 2)$	**c** $2(x + 5) = 4(x - 2)$		**d** $x = 9$	**e** 28 square units		
9	**a** 11	**b** 12	**c** 4	**d** 8				
	e 13	**f** 18	**g** 23	**h** 32	**i** 28			
10	**a** $a + 6$	**b** $\frac{1}{4}(x + 6) = 12$		**c** 18 years old				

Challenge: Three shapes

30 units

Lesson 15.4 Rearranging formulae

Learning objective

- To change the subject of a formula

Links to other subjects

- **Science** – to rearrange equations such as Ohm's law and mass density volume
- **Business studies** – to make financial calculations

Resources and homework

- Pupil Book 2.2, pages 270–273
- Homework Book 2, section 15.4
- Online homework 15.4, questions 1–10

Key word

- subject

Problem solving and reasoning help

- The financial skills activity at the end of Exercise 15D in the Pupil Book involves substituting and rearranging the simple interest formula. Pupils who have understood the lesson should be able to tackle this with minimal input. Challenge **more able** pupils by adding time into the equation or by being introduced to the compound interest formula.

Common misconceptions and remediation

- Pupils will sometimes forget to apply BIDMAS correctly when rearranging formulae and hence get an incorrect answer. Remind pupils that when rearranging a formula, they should simply follow the methods learned earlier in the chapter for solving equations. Pupils who struggle to rearrange formulae may find the input-output method easier to understand.

Probing questions

- Explain how you would rearrange this formula. What would you do first? Why? What next?
- What parts are the same and what parts are different when you rearrange a formula or solve an equation?

Part 1

- This puzzle is best thought about individually: A man was buying balloons. He said: 'Some 20p balloons, six times as many 10p balloons and make up the rest in 25p balloons.' He handed over £6 and said: 'No change please.'
- How did the shopkeeper sort out the number of balloons?
- Ask pupils to find the solution, reminding them that there must be no change and no overcharging. A strategy to help find the solution using algebra is to let the number of 20p balloons be x. Then you have $6x$ balloons at 10p and the rest at 25p.
- The cost of the 20p balloons is $20x$, and the cost of the 10p balloons is $60x$. So the cost of both together is $80x$.
- The remainder of the £6 must go on 25p balloons, so to generate no change, $80x$ must be a multiple of 25p. Going through the possibilities of $x = 1$, $x = 2$, $x = 3$, …, it is found that $x = 5$ gives $80x = 400$, which is a multiple of 25p.
- Therefore the solution is five 20p balloons, thirty 10p balloons and eight 25p balloons.
- Go through the algebraic solution with the class if there is time.

Part 2

- Ask if anyone can remember what a *formula* is.
- Draw from pupils that a formula is an equation for working out a value from known facts.
- Explain that *Celsius* (also called *centigrade*) and *Fahrenheit* are two different scales for measuring temperature. The former has largely replaced the latter in the UK. Introduce the notation °C and °F and how these are pronounced.

Show the formula for converting a temperature C in degrees Celsius into a temperature F in degrees Fahrenheit: $F = \frac{9C}{5} + 32$

- Use the formula to convert 35 °C into degrees Fahrenheit:
  $C = 35$, so $F = \frac{9 \times 35}{5} + 32 = 63 + 32 = 95$
- Ask pupils if anyone can think of a formula that uses two *variables* to work out a third, for example, the 'area of a triangle' formula. If there are no other suggestions, use the triangle formula. Illustrate how to substitute into the formula to calculate one value from two others.
- The area of a triangle is given by $A = \frac{1}{2}bh$, where b is the base length and h is the vertical height of the triangle. The area of a triangle with a base length of 5 cm and a vertical height of 8 cm, is given by: $A = \frac{1}{2} \times 5 \times 8 = 20\text{ cm}^2$
- Do another example with this formula, but this time the area and base length are known, and the height must be calculated. For example, $A = 36\text{ cm}^2$ and $b = 8$ cm gives $h = 9$ cm. Substitute in the known values *before* rearranging to find the height.
- **Pupils can now do Exercise 15D from Pupil Book 2.2.**

Part 3

- Ask pupils to give you as many formulae to do with perimeter, area, surface area and volume as they can think of. Write these on the board. Now ask pupils to rearrange them.

Answers

Exercise 15D

1 **a** 70 **b** 40 **c** 40 **d** 40

2 **a** $t = 5 - 2s$ **b** $t = w + 6.5$ **c** $t = \frac{a}{4}$ **d** $t = 2m$

3 **a** $n = T - m$ **b** $n = q - t + 12$ **c** $n = \frac{1}{3}(y - a)$ **d** $n = \frac{y}{3} - 1$

4 **a** 8 **b** 131 **c** $2m = x + y$, $x = 2m - y$
d 34 **e** $y = 2m - x$ **f** 10.7

5 **a** $u = v - 10t$ **b** 11.4 **c** $t = \frac{1}{10}(v - u)$ **d** 2.4

6 **a i** 58 **ii** 110 **b** $x = \frac{1}{4}(a - 30)$
c i 5 **ii** 10

7 **a** $x = \frac{1}{5}(y - 12)$ **b** $x = 3(y + 2)$ **c** $x = 50 - y$
d $x = \frac{1}{25}(y - 45)$ **e** $x = 20 - 3y$ **f** $x = \frac{1}{4}(18 - y)$

8 **a** $a = k - 3b + 1$ **b** $b = \frac{1}{3(k - a + 1)}$

9 **a** 25 **b** $t = \frac{v - u}{a}$ **c** 6

10 **a** $a = \frac{1}{2}P - b$ **b** $b = \frac{1}{2}P - a$ **c** $a = \frac{A}{b}$ **d** $b = \frac{A}{a}$

11 **a** The perimeter is the sum of the three sides
b 38 **c** $y = p - 2x$ **d** $x = \frac{1}{2(p - y)}$

12 **a** $p = \frac{1}{10}(t - 20q)$ **b** $q = \frac{1}{10}(t - 10p)$

Financial skills: Paying interest

A **a** £72 **b** £472
B **a** £60 **b** £2060
C **a** $R = \frac{100}{A}$ **b** 16%

Review questions

(Pupil Book pages 274–275)

- The review questions will help to determine pupils' abilities with regard to the material within Chapter 15.
- These questions also draw on the maths covered in earlier chapters of the book to encourage pupils to make links between different topics.
- The answers are on the next page of this Teacher Book.

Mathematical reasoning – Using graphs to solve equations

(Pupil Book pages 276–277)

- In this activity pupils use mathematical reasoning to make links between equations and formula and their graphical representation. By comparing graphical and algebraic representations pupils check their ability to solve equations. This ability to use different representations to check their understanding is a valuable generic skill.
- In order to tackle the questions in this activity, pupils will need to be comfortable solving the equations encountered in this chapter, and how to draw graphs from equations of straight lines. Model some examples on the board before pupils work on questions 1 and 2.
- Make sure that pupils draw the axes accurately and label the scales correctly.

Answers to Review questions

1 **a** 4.5 **b** 21 **c** 45 **d** 3.5

2 **a** 8 **b** 17 **c** 6 **d** 4.5

3 **a** 22 **b** 10 **c** 20 **d** 34

4 **a** 18 **b** 20 **c** 23 **d** 8

5 **a** $x + x + 18 + 2(x - 5) = 180$ **b** 43°, 61° and 76°

6 **a** A is $4(4 + a)$; B is $6a$ **b** $4(4 + a) = 6a$

c $a = 8$; area = 48 **d** No, perimeters are 32 and 28

7 **a** $b = \frac{2A}{h}$ **b** $h = \frac{2A}{b}$

8 $c = \frac{f - 32}{1.8}$

9 $x = 2(y - 5)$

10 **a** $a = \frac{p - 4b}{2}$ **b** $b = \frac{p - 2a}{4}$

11 **a** $a = 2m - b$ **b** $b = 2m - a$

Answers to Reasoning – Using graphs to solve equations

1 **a** $x = 2$

b Middle row: 2, 4, 6, 8, 10, 12; bottom row: 8, 7, 6, 5, 4, 3

c–e

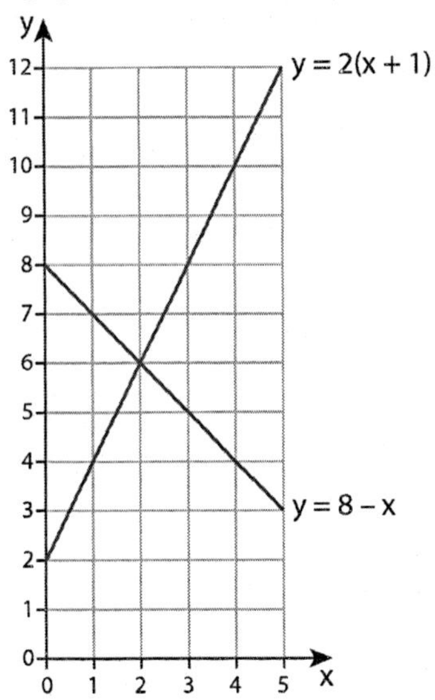

2 **i a** $x = 1$ **b** $x = 8$ **c** $x = 2$ **d** $x = 7.5$

ii Pupils' check own answers

16 Comparing data

Learning objectives

- How to construct grouped frequency tables for data
- How to interpret and draw grouped frequency diagrams
- How to compare two distributions by using an average and the range
- How to select the correct average when analysing data

Prior knowledge

- How to construct a frequency table from given data
- How to find the mode, median, range and modal class for grouped data
- How to calculate the mean from a set of data
- How to construct graphs and diagrams to represent data

Context

- How many people are there in the world? Or even in our country? How do these people live? What do they eat and drink? How big are their families?
- Encourage pupils to consider how statistics are used to explore questions like this. For example, a census is a huge survey that is used to find out information about each man, woman and child in a country. Pupils need to consider how to present this type of information, or data.
- Pupils also need to think about how we use statistics to model populations where it is difficult, or in many cases impossible, to gather all the population information.

Discussion points

- Make up a statement or question for this chart or graph using one or more of these words:
  - total
  - range
  - mode.

Associated Collins ICT resources

- Chapter 16 interactive activities on Collins Connect online platform
- *Displaying data* and *From pastime to sport* Wonder of Maths on Collins Connect online platform

Curriculum references

Statistics

- Describe, interpret and compare observed distributions of a single variable through: appropriate graphical representation, and appropriate measures of central tendency and spread
- Construct and interpret appropriate tables, charts, and diagrams, including pie charts

Reason mathematically

- Explore what can and cannot be inferred in statistical and probabilistic settings, and begin to express their arguments formally

Solve problems

- Begin to model situations mathematically and express the results using a range of formal mathematical representations

Fast-track for classes following a 2-year scheme of work

- Use one or two examples to check pupils' understanding from Lesson 16.1 and Lesson 16.2. If pupils are fluent and confident with the concepts, move straight to Lesson 16.3 and Lesson 16.4, where pupils will compare data and make decisions about the most appropriate statistical measures they should use.

Lesson 16.1 Grouped frequency tables

Learning objective

- To create a grouped frequency table from raw data

Resources and homework

- Pupil Book 2.2, pages 279–281
- Intervention Workbook 2, pages 76–79
- Homework Book 2, section 16.1
- Online homework 16.1, questions 1–10

Links to other subjects

- **Geography** – to analyse population data
- **History** – to compare historical data

Key words

- class
- continuous
- discrete
- frequency
- grouped frequency table

Problem solving and reasoning help

- **PS** questions 5 and 6 of Exercise 16A in the Pupil Book give pupils the opportunity to see statistics being used in real-life situations with which they may be familiar. Encourage pupils to discuss situations when they think they might be able to use grouped data in real life. Suggest that they talk to family and friends to see if they use data in their everyday lives. Help pupils to see that data can be used in life, learning and earning.

Common misconceptions and remediation

- Pupils often struggle to identify appropriate groups because they want to follow a set of rules. This makes it difficult for them to make links between the data and the question that is being asked. As a result, pupils may fail to grasp how the links should inform their decisions. Give pupils lots of real-life examples to discuss. This will help them to see that though there may not be a correct answer, some groupings are more efficient or informative than others.

Probing questions

- What is discrete/continuous data? Give me some examples.
- How do you go about choosing class intervals when grouping data for a bar chart or frequency diagram?

Part 1

- This section is the introduction to Part 2. Write 0–10, 10–20 and 20–30 on the board or OHP. Tell the class that these represent time intervals for how long pupils spend on pieces of work. Say that there is a problem with using these intervals. What is the problem? Pupils may point out the overlapping time intervals and that you cannot record for more than 30.
- Now change the intervals to 0–10, 11–20, 21–30 and 31+ and ask pupils if this solves the problem. Hopefully pupils will spot that there are now gaps between the classes.
- Ask the class for ideas on how to record the times without these problems. Prompt pupils to use the words *inclusive* and *exclusive*.
- Finally, write $0 < \text{time} \leq 10$ on the board and ask pupils to explain what this means. This should lead to the use of this notation for class intervals. Compare this with the meaning of $0 \leq \text{time} < 10$ and point out that it is important to check carefully which is being used.

Part 2

- Copy the following table onto the board or an OHP:

Height of book, h (cm)	Frequency
$22 < h \le 24$	
$24 < h \le 26$	
$26 < h \le 28$	
$28 < h \le 30$	

- Ask pupils to measure the heights of different books. Record their responses. You may wish to keep a tally alongside the table. Emphasise that the symbol ≤ could be the first or second relation, but that this should be kept consistent. Explain to the group again that, for example, 24 will be recorded in the first class.
- Complete the frequency table.
- **Pupils can now do Exercise 16A from Pupil Book 2.2.**

Part 3

- Discuss how to collect a large set of data. Talk about the advantages of working in groups.
- Explain to pupils that the next step is to collect large sets of data and for them to decide on the size of the class interval.
- Put a frequency table of ages on the board and say that you wish to look at how the population in the area is made up.
- Use the classes of: 0 < age ≤ 10, 10 < age ≤ 20, 20 < age ≤ 40 and age > 40.
- Ask pupils why sometimes it might be appropriate to use unequal class intervals such as these. Talk about small sets of data, for example, age > 90.

Answers

Exercise 16A

1 **a** $10 < T \le 20$ **b** $0 < T \le 12$

2 **a** Frequencies are 2, 4, 6, 6, 2 **b** $1.70 < h \le 1.80$

3 **a** Frequencies are 4, 2, 1, 3, 2, 2 **b** $0 < M \le 1$ **c** $\frac{3}{14}$

4 **a** Frequencies are 3, 5, 3, 3, 2 **b** $10 < T \le 12$ **c** 9°

5 $35 < P \le 40$

6 **a** Dr Speed and Dr Bell did

b For example Dr Speed good range, Dr Bell too many long consultations with none under five minutes, Dr Khan quicker with none over ten minutes and nearly twice as many under five minutes as between five and ten.

Activity: Textbook text

Answers will vary.

Lesson 16.2 Drawing frequency diagrams

Learning objectives

- To interpret frequency diagrams
- To draw a frequency diagram from a grouped frequency table

Resources and homework

- Pupil Book 2.2, pages 281–284
- Intervention Workbook 2, pages 76–79
- Homework Book 2, section 16.2
- Online homework 16.2, questions 1–10

Links to other subjects

- **Geography** – to analyse population data
- **History** – to compare frequency of historical events in different countries or periods

Key words

- block graph
- frequency diagram
- line graph

Problem solving and reasoning help

- **PS** question 3 of Exercise 16B in the Pupil Book and the activity at the end of the exercise are both an opportunity to explore with pupils the misconception outlined below. Pupils could role-play their solution, with the rest of the class providing feedback on their diagrams and brochures. Feedback should be based on a set of agreed criteria such as: accuracy, ease of use of diagrams, did diagrams enhance explanation, suitability for target audience.

Common misconceptions and remediation

- Pupils often do not associate drawing frequency diagrams with the analytical process of comparing outcomes or gathering evidence to support a hypothesis. Give pupils a range of examples, good and less good, from real-life contexts. Ask them to discuss how useful the representations are in terms of assessing the evidence and/or delivering a clear message.

Probing questions

- What is important when choosing the scale for the frequency axis?
- When considering a range of graphs representing the same data, think about:
  - Which is the easiest to interpret? Explain why.
  - Is this graphical representation helpful in addressing the hypothesis? If not, why and what would you change?

Part 1

- Write the words *frequency*, *table*, *collect*, *tally*, *diagram* and *data* on the board.
- Ask pupils to sort the words into an order that they can explain, for example: *collect, data, tally, frequency, table, diagram.*
- Now ask pupils to make the words into complete sentences. For example: 'I am going to collect some data on a tally chart. I will then set them out as a frequency table and draw a frequency diagram.'
- You can add other words such as: *compare*, *statistics*, *continuous* and *discrete*.
- You could use this activity to establish a set order of working with statistics, and as part of the literacy strategy. A further step could be to ask pupils to spell some of the words before writing them on the board.

Part 2

- Sketch a *bar chart* on the board or OHP. Leave gaps between the bars and label the bars, for example: cats, dogs, ...

- Sketch a *frequency diagram* (with no gaps and equal class intervals) on the board or OHP, labelling the divisions between the bars with *continuous data*.
- Ask pupils to name the differences between the diagrams. Pupils may point out the gaps. Ask why there are gaps on a bar chart. Lead pupils into talking about *discrete data*.
- Refer again to the frequency diagram and point out, if necessary, that this diagram has continuous data. State that there should be no gaps.
- Write a checklist of what is needed for a good frequency diagram for continuous data:
  - a title
  - suitable class intervals
  - axes labelled, with the horizontal axis labelled at the class boundaries
  - neat ruled bars
  - no gaps
- Tell the class that, as well as this type of frequency diagram, they will also be required to draw and read from time-series graphs. Refer to Examples 3 and 4 on pages 282 and 283 of the Pupil Book and ask pupils to give you some facts from the graphs. Point out that these types of graphs can be used to look at trends.
- **Pupils can now do Exercise 16B from Pupil Book 2.2.**

Part 3

- Write four headings on the board: 'Time', 'Temperature', 'Length', 'Mass'.
- Tell the class that these are the most often used categories for continuous data.
- Ask the class to give you units to write in the four columns. Write down the units as pupils give them to you, for example: seconds, °C, metres, kilograms.
- Stress that continuous data have to have a continuous scale on diagrams.

Answers

Exercise 16B

1 Check pupils' diagrams

2 **a** City A **b** City B **c** 10 **d** 5°

3 Check pupils' diagrams

Activity: Comparing holiday destinations

Answers will vary.

Lesson 16.3 Comparing data

Learning objective

- To use mean and range to compare data from two sources

Links to other subjects

- **Physical Education** – to compare the performances of different teams or individuals over time; use this to plan future training schedules
- **Geography** – to compare the frequency of climate events over time

Resources and homework

- Pupil Book 2.2, pages 285–286
- Intervention Workbook 1, pages 60–63
- Homework Book 2, section 16.3

Key words

- No new key words for this topic

Problem solving and reasoning help

- **MR** question 4 gives pupils the opportunity to use a statistical measure of central tendency (mean) and spread (range) to compare data in a familiar context. It may be useful to discuss the different roles these measures play. Pupils could suggest how the different statistical measures might be used in other real-life situations with which they might be familiar. Pupils could also discuss examples from the media including the purpose of using each measure and whether they think it is a reasonable use of the measure being used. This is extended to a more complex example in the activity at the end of Exercise 16C in the Pupil Book.

Common misconceptions and remediation

- Pupils may struggle with interpreting data accurately. Provide lots of opportunities for pupils to compare different representations. Encourage pupils to generate their own questions about the data and to assess the advantages and disadvantages of different representations and statistical measures.

Probing questions

- How will you make sense of the data you have collected?
- What options do you have to organise the data?
- What other questions could you ask of the data?

Part 1

- Use a target board as shown:

1	3	5	7	9	11	13
2	4	6	8	10	12	14

- Ask the class to look at the top row and tell you the *mean* (or average) value. Ask pupils how they worked it out. Prompt them to refer to the pattern and the symmetry of the row.
- Now ask the class for the mean of the second row, and ask pupils to *compare* the first and second rows. Ask them what has happened to each number.
- Now ask pupils for the *range* of the numbers in the top row and the range in the bottom row.
- Establish that, although the mean has increased by 1, the *range* is unchanged.
- Repeat, using different second rows, for example: 2, 6, 10, 14, 18, 22, 26.

Part 2

- Write on the board: 'Global warming – average temperatures will rise by two degrees over next 100 years.' Ask the class to explain what will happen to the lowest temperature, the highest temperature and the range.
- Point out that warming is unlikely to happen uniformly. Ask pupils to describe mathematically, the effect of, for example, the lowest temperature falling and the highest temperature rising. Use terminology such as *variation* and *consistency*.
- Now introduce the idea of comparing means and ranges. Draw a table on the board giving the number of goals by football teams A and B in six different matches:

	A	B
	1	0
	0	3
	1	0
	1	7
	1	2
	2	0
Mean	1	2
Range	2	7

- Discuss the differences. Emphasise that pupils must compare, and not just repeat the data given in the question. Which team would pupils support?
- **Pupils can now do Exercise 16C from Pupil Book 2.2.**

Part 3

- Ask the class to explain the advantage of a golf player having a small range for the number of shots taken per hole. Talk about variation and consistency.
- You may wish to give the class some data for five holes. For example, player A hits 4, 4, 4, 5, 4 and player B hits 3, 8, 5, 4, 1. Both have the same mean of 4.2 – but look at the ranges.
- Now talk about different pupils spending considerably different lengths of time on their homework. Talk about different average times and less consistency.

Answers

Exercise 16C

1 **a** 17 min **b** 16.4 min

2 **a** 10, 11, 12.5, 13.5
b For example, the further north the county was, the colder and smaller the range.

3 Matt scores consistently well, where Jon's scores vary a lot.

4 **a** Everlast 6 hrs, 2; Powercell 4.2 hrs, 3; Electro 8.8 hrs, 1
b For example Electro – lasts the longest and very reliable. Powercell – cheap, just a third of the cost of Electro but last about half as long.

5 Cardiff 13.6, 14; Edinburgh 11.8,12; London 13.8, 16. Edinburgh is generally the coldest, with London being slightly warmer than Cardiff but a more variable temperature.

Activity: Comparing populations

Answers will vary.

Lesson 16.4 Which average to use?

Learning objective

- To understand when each different type of average is most useful

Links to other subjects

- **Science** – to compare biological population data
- **Religious education** – to compare social data on religious beliefs

Resources and homework

- Pupil Book 2.2, pages 287–289
- Homework Book 2, section 16.4

Key words

- No new key words for this topic

Problem solving and reasoning help

- Before tackling this lesson it would be very useful to work through the table on page 287 of the Pupil Book to explain the advantages and disadvantages of different measures of central tendency and range. Engage pupils in a conversation about why we use these measures in real life and ask for some further examples of their own. The **MR** and **PS** questions in Exercise 16D of the Pupil Book give pupils the opportunity to apply their understanding to recognisable, though possibly less familiar real-life contexts. You could extend this to the more complex investigation at the end of the exercise. You may want to do the activity collaboratively with the class or model an example first. With pupils, consider carefully what type of, and how much data you will collect and how to identify possible sources for the data.

Common misconceptions and remediation

- The non-technical use of the language of central tendency and spread in real life and in the media often confuses pupils. Tackle this confusion by using clear examples. Some pupils may not grasp the difference between 'estimating the mean of ...' and 'calculating an estimate of the mean of ...'.

Probing questions

- Why is it only possible to estimate the mean (median, range) from grouped data?
- Talk me through the steps you take to estimate the median from grouped data.
- Explain to me when the estimated mean is not the most appropriate average, that is, when is it not a representative value?
- How would you decide whether to, for example, buy 'type X' instead of 'type Y'?

Part 1

- Write £20 000 on the board. Say that you have seen a brand new car, which costs £20 000.
- Now use 'show me' cards and ask pupils to write on their card what they think the value of the car will be in one year's time.
- Identify the smallest and biggest answers. Ask a pupil to tell you the range of answers.
- Ask a pupil to explain, for example, why she or he said £15 000.
- Now repeat, but this time pupils should work in pairs and agree on a price between them.
- Again look at the range of answers. Hopefully the value will be less than before.
- Ask a pupil to explain why the range is now less.
- Repeat this exercise for different items, for example, a tin of beans costing 35p now.

Part 2

- Display a set of data suitable for comparison, for example: the values of used cars at different showrooms or attendance figures at different sports event.
- Ask questions such as: 'Where should you buy your car?', 'Which sport is most popular?' Discuss the use of different statistical measures. Use the table on page 287 of the Pupil Book to support the discussion. Try to use examples that suggest the use of different statistical measures. For **more able** pupils, you could draw more complex, less familiar data sets from the Gapminder website: **http://www.gapminder.org/**
- **Pupils can now do Exercise 16D from Pupil Book 2.2.**

Part 3

- Tell pupils to imagine that they are the managing director of a company that wants to move out of London to another part of the country.
- Ask them to suggest ways to use statistical information to help them make their decision.
- Pupils should think about what data they would need, how they might collect it and what statistical measures they might use.
- You could give **less able** pupils a profile of the company and ask them to choose from a given set of measures. Ask **more able** pupils to write their own profile and perhaps compare the requirements of two companies.
- Working in pairs or small groups, ask pupils to find specific examples to share with the class.
- Depending on the group, they could use alternative scenarios or contexts.

Answers

Exercise 16D

1 **a i** 5.8 **ii** yes **b i** 2 **ii** yes **c i** 8 **ii** yes
d i 10 **ii** no **e i** 2 **ii** no **f i** 8.7 **ii** no

2 **a** $14 < T \le 16$ **b** No individual mode

3 **a** 9, yes **b** 9, no, majority 10 **c** 9, no, ten is an outlier
d 9, yes **e** 9, yes **f** 12, yes

4 **a** Frequencies are 7, 2, 5, 5, 3, 6 **b** $0 < M \le 1$ **c** \$2.90
d The median is a single value, so it is better to choose the modal class

5 **a** Mode **b** Mean
c i Mode as that is what most workers will be paid
ii The mean as it takes their higher wages into account also
iii The median as it shows middle of the range but skewed to the bottom end

6 Craig 5.3,5 ; Len 7.1,2 ; Darcy 6.6,3 ; Bruno 7.4,3
Craig is the lowest scoring but has the largest range of scores.
Bruno is the highest scorer with the same range as Darcy. Len's scores vary the least.

7 **a** Highest average score **b** Can score higher than Joe
c Joe as he is more consistent and a higher mean

Review questions

(Pupil Book pages 290–291)

- The review questions will help to determine pupils' abilities with regard to the material within Chapter 16.
- These questions also draw on the maths covered in earlier chapters of the book to encourage pupils to make links between different topics.
- The answers are on the next page of this Teacher Book.

Problem solving – Technology questionnaire

(Pupil Book pages 292–293)

- This activity is designed to combine all the lessons in this chapter and take pupils sequentially through the steps of tabulating and displaying data in a familiar real-life problem context.
- All the data is given but pupils will need to read the think carefully about how they display it so they can make valid comparisons.
- Ask pupils further questions as a warm-up to get them thinking about the purpose of the activity. Here are some examples:
    - How many music downloaded tracks do you have?
    - Do you think you have more or less than most people your age?
    - How many CDs do you possess?
    - Do your parents use CDs or download music? What about other family members?
    - Which is your favourite Wii sporting game?
- Pupils can now work on questions 1 to 5 individually or in groups.
- You could ask pupils to develop this topic further by using the internet to research other data linked to the example questions above.

Answers to Review questions

1 **a** No, no numbers to calculate with
b Yes, Lady Gaga appears more times than any other
2 **a** 7.5 **b** For example 3, 3 and 6 **c** For example 1, 4, 6, 7, 7
3 **a** Nothing **b** Nothing **c** 796 kg
4 **a** Check pupils' graphs **b** **i** £35 **ii** 21cm
5 **a** $10 < T \leq 15$ **b** $5 < T \leq 10$ **c** $\frac{13}{20}$
6 **a** Gary highest mean score of 45.5 **b** Mark has the two highest scores
c Gary, as he is the more consistent with the highest mean
7 **a** Check pupils' tables
b Check pupils' diagrams **c** 16 – 20
8 **a** 5 **b** ≥7 **c** 3
9 **a** 17 cm **b** 12.5 cm^2

Answers to Problem solving – Technology questionnaire

1 Frequencies should be as follows:

	Boys	Girls
Baseball	1	0
Bowling	4	19
Boxing	13	0
Golf	5	4
Tennis	2	2

2 Pie charts showing frequencies from question 1
3 Check that the class sizes used are sensible
4 Frequency chart to match pupils' tables from question 3
5 **a** $\frac{1}{2}$ **b** $\frac{27}{50}$ **c** $\frac{27}{50}$ **d** $\frac{2}{25}$

1 Working with numbers

Learning checklist

I can find square and cube numbers and square and cube roots. ☐
I can use a calculator to work out powers of numbers. ☐
I can find common factors for pairs of numbers. ☐

I can multiply and divide negative numbers, for example, $-5 \times 3 = -15$. ☐
I know that the square roots of positive numbers can have two values, one positive and one negative. ☐
I can find the lowest common multiple (LCM) for pairs of numbers. ☐
I can find the highest common factor (HCF) for pairs of numbers. ☐
I can write a number as the product of its prime factors. ☐

2 Geometry

Learning checklist

I can rotate a simple shape about a point. ☐
I can translate a shape. ☐

I know the angle properties of parallel lines. ☐
I can classify the different types of quadrilateral. ☐
I know the geometric properties of quadrilaterals. ☐
I can rotate a shape about a given centre of rotation on a coordinate grid. ☐
I can construct perpendicular lines and bisect angles. ☐

3 Probability

Learning checklist

I can calculate probabilities involving equally likely outcomes. ☐
I can calculate probability from experimental data. ☐

I know what mutually exclusive outcomes are. ☐
I can use the probability of an outcome to calculate the probability that the outcome does not happen. ☐
I can use sample spaces and Venn diagrams to help calculate probabilities. ☐

4 Percentages

Learning checklist

I can write one value as a percentage of another value. ☐
I can use percentages to compare two quantities. ☐
I can use a multiplier to calculate a percentage increase or decrease on a value. ☐
I can write a change of value as a percentage increase or decrease. ☐

5 Sequences

Learning checklist

I can use flow diagrams to generate sequences. ☐
I know how to work out the nth term of a sequence when given the rule. ☐
I can write out the terms of the Fibonacci sequence. ☐

I can work out the rule for the nth term of a sequence. ☐

6 Area of 2D and 3D shapes

Learning checklist

I can use the appropriate formulae to find the area of triangles, parallelograms and trapezia. ☐

I can use the formula $S = 2lw + 2lh + 2wh$ to work out the surface area of a cuboid. ☐

7 Graphs

Learning checklist

- I can complete a table of values for a linear equation and use this to draw a graph of the equation. ☐
- I can calculate the gradient of a straight line drawn on a coordinate grid. ☐
- I can work out an equation of the form $y = mx + c$ from its graph. ☐
- I can complete a table of values for a simple quadratic equation and use this to draw a graph of the equation. ☐
- I can draw and interpret graphs that illustrate real-life situations. ☐

8 Simplifying numbers

Learning checklist

	I can multiply and divide numbers by powers of ten.	☐
	I can round large numbers.	☐
	I can round numbers to a specific number of significant figures.	☐
	I can write a large number in standard form.	☐
	I can multiply numbers in standard form.	☐

9 Interpreting data

Learning checklist

I can interpret a pie chart by using the angles made by sectors at the centre of the chart. ☐

I can use the scaling method to construct a pie chart. ☐
I can understand a scatter graph. ☐
I can draw a scatter graph from given data. ☐
I can describe any correlation shown in a scatter graph. ☐

10 Algebra

Learning checklist

I can simplify algebraic expressions involving the four basic operations. ☐
I can simplify algebraic expressions with several terms by combining like terms. ☐

I can expand an expression by removing brackets. ☐
I can manipulate algebraic expressions. ☐
I can identify equivalent expressions. ☐
I can write algebraic expressions involving powers. ☐

11 Congruence and scaling

Learning checklist

I can recognise congruent shapes. ☐
I know how to make a scale drawing. ☐

I can enlarge a 2D shape by a scale factor about a centre of enlargement. ☐
I can use ratio to compare lengths, areas and volumes of 2D and 3D shapes. ☐
I can use map ratios. ☐

I can recognise congruent triangles. ☐

12 Fractions and decimals

Learning checklist

- I can add and subtract mixed numbers. ☐
- I can multiply a fraction or a mixed number by an integer. ☐
- I can divide an integer by a unit fraction or a unit fraction by an integer. ☐
- I can multiply simple large or small decimal numbers without using a calculator. ☐
- I can divide simple large or small decimal numbers without using a calculator. ☐

13 Proportion

Learning checklist

I can recognise and use formulae for two variables that are in direct proportion. ☐

I can recognise and draw a graph for two variables that are in direct proportion. ☐

I can work out a formula connecting two variables that are in direct proportion. ☐

I can work out a formula connecting two variables that are in inverse proportion. ☐

I can recognise and draw a graph for two variables that are in inverse proportion. ☐

I can decide whether two variables are in direct proportion or inverse proportion. ☐

14 Circles

Learning checklist

I can use the formula $C = \pi d$ when calculating the circumference of a circle. ☐

I can use the formula $A = \pi r2$ when calculating the area of a circle. ☐

15 Equations and formulae

Learning checklist

I can solve equations involving brackets and fractions. ☐
I can solve equations where the variable occurs on both sides. ☐

I can change the subject of a formula in simple cases. ☐

16 Comparing data

Learning checklist

I can compare the range and the mean from two sets of data. ☐
I can decide which average is the best to use in different circumstances. ☐

I can construct grouped frequency tables. ☐
I can construct and interpret grouped frequency diagrams. ☐

3-year scheme of work

The following scheme of work provides a suggestion for how Pupil Book 2.2 can be taught over the course of one year, as part of a 3-year Key Stage 3 course.

Please note that you can recombine the test questions provided on Collins Connect to create new tests if your frequency of assessment differs from that below, or if you wish to combine content from different chapters in your own half-term tests.

This scheme of work is provided in editable Word and Excel format on the CD-ROM accompanying this Teacher Pack.

Chapter	Lesson	No. of hours	Learning objective	Comments/ suggestions
Half-term / Term 1				
1 Working with numbers	1.1 Multiplying and dividing negative numbers	1	• To carry out multiplications and divisions involving negative numbers.	One of the main misconceptions when multiplying two negative numbers together is consistently giving a negative answer. Another problem pupils have when multiplying two numbers together is that they often think the sign of the answer is determined by the sign of the largest number. Make sure that pupils do not rush through their work and that they have a clear understanding of the rules.
	1.2 Factors and highest common factors (HCF)	1	• To understand and use highest common factors	Students sometimes confuse factors and multiples. (Tell them that multiples come from multiplying.)
	1.3 Lowest common multiples (LCM)	1	• To understand and use lowest common multiples	
	1.4 Powers and roots	2	• To understand and use powers and roots	
	1.5 Prime factors	1	• To understand what prime numbers are • To find the prime numbers of an integer	
	Challenge – Blackpool Tower	1		This activity is designed to give pupils the opportunity to apply their learning to a real-life multi-step problem.
2 Geometry	2.1 Angles in parallel lines	1	• To calculate angles in parallel lines	
	2.2 The geometric properties of quadrilaterals	1	• To know the geometric properties of quadrilaterals	
	2.3 Rotations	1	• To understand how to rotate a shape	Pupils struggle to visualise transformations Give them plenty of practice and if possible use active geometry packages such as Geogebra to help them **http://www.geogebra.org/cms/en/** You could also use readymade examples on Geogebra tube **http://www.geogebratube.org/material/show/id/2163**
	2.4 Translations	1	• To understand how to translate a shape	
	2.5 Constructions	1	• To construct the mid-point and the perpendicular bisector of a line • To construct an angle bisector	Pupils are often not precise enough when doing constructions in mathematics. Give them the opportunity to assess the errors in exemplars and explain how they can be avoided. Use dynamic geometry software to support learners.
	Challenge – More constructions	1		This challenge gives pupils the opportunity to extend their learning to more complex constructions. They need to be able to reproduce a set of instruction that extend what they have

				already done in the lesson.
Chapter 1–2 assessment on Collins Connect				
3 Probability	3.1 Probability scales	1	• To use a probability scale to represent a chance	This chapter builds on previous knowledge of probability and extends this first to see how probability is applied differently to theory and experiments, and then to being able to compare the two results critically.
	3.2 Mutually exclusive events	1	• To recognise mutually exclusive events	
	3.3 Using a sample space to calculate probabilities	1	• To use sample spaces to calculate probabilities	
	3.4 Experimental probability	2	• To calculate probabilities from experiments	
	Financial skills – Fun in the fairground	1		In this activity learners extend their understanding of probability to a common real-life application that they may not have previously considered. It also makes a real-life link between probability and financial skills.
Half-term				
Half-term / Term 2				
4 Percentages	4.1 Calculating percentages	1	• To write one quantity as a percentage of another • To use percentages to compare quantities	Fractions, decimals and percentages are everywhere in real life and it is important for confidence and accuracy to be able to move between these different representations. This chapter reinforces the links between fractions, decimals and percentages.
	4.2 Calculating percentage increases and decreases	2	• To use a multiplier to calculate a percentage change	
	4.3 Calculating a change as a percentage	2	• To work out a change in value as a percentage increase or decrease	
	Challenge – Changes in population	1		This activity is designed to give pupils the opportunity to demonstrate their understanding of percentage change to a real-life situation. All the information they need is provided but they will need to read the questions carefully to decide which information they need and what mathematical skills to use.
5 Sequences	5.1 Using flow diagrams to generate sequences	1	• To use flow diagrams to generate sequences	The ability to generalise is crucial in a complex modern society. Being able to identify and generate number sequences is the first step towards progressing from the particular to the general in mathematics.
	5.2 The *n*th term of a sequence	2	• To use the *n*th term of a sequence	
	5.3 Working out the *n*th term of a sequence	2	• To work out the *n*th term of a sequence	
	5.4 The Fibonacci sequence	1	• To know and understand the Fibonacci sequence	
	Investigation – Pond borders	1		Pupils apply their understanding of sequences to a real-life scenario. They will need to work methodically and be able to justify their solutions. Ask more able pupils to generalise their rules for an $m \times n$ pool.
Chapter 3–5 assessment on Collins Connect				
6 Area of 2D and 3D shapes	6.1 Area of a triangle	1	• To work out the area of a triangle	Pupils should understand that the height of a triangle, parallelogram and trapezium (except in some specific examples) is the vertical height, not the length of a side. Encourage pupils to see how they can use what they already know, for example, the area of a triangle and a rectangle, to work out things they may not know or have forgotten.
	6.2 Area of a parallelogram	1	• To work out the area of a parallelogram	
	6.3 Area of a trapezium	1	• To work out the area of a trapezium	

	6.4 Surface areas of cubes and cuboids	2	• To find the surface areas of cubes and cuboids	Pupils often confuse the concept of surface area and volume. Use concrete examples to help them understand the difference.
	Investigation – A cube investigation	2		Pupils apply their understanding of area to a more complex problem. They will need to work methodically and be able to explain their solutions. Ask more able pupils to justify any rules by revisiting the structure of the problem.
Holidays				
Half-term / Term 3				
7 Graphs	7.1 Graphs from linear equations	1	• To recognise and draw the graph of a linear equation	This chapter builds on previous work on mapping diagrams and graphs covered in Year 7. The important concept of the gradient of a straight line is introduced in this chapter and the form $y = mx + c$ for a straight line is explored.
	7.2 Gradient (steepness) of a straight line	1	• To work out the gradient in a graph from a linear equation • To work out an equation of the form $y = mx + c$ from the graph	
	7.3 Graphs from simple quadratic equations	2	• To recognise and draw the graph from a simple quadratic equation	
	7.4 Real-life graphs	2	• To draw graphs from real-life situations to illustrate the relationship between two variables	
	Challenge – The M25	1		A common response to algebra is to ask how it can be used. This activity provides an everyday use of algebra in terms of graphical representation of algebraic relationships set in real life contexts. Encourage pupils to suggest possible questions.
8 Simplifying numbers	8.1 Powers of 10	1	• How to multiply and divide by powers of 10	This chapter builds on previous work with decimals, introducing powers of 10 as a lead in to working with standard index form. Estimation is used as a means of teaching whether answers are realistic or sensible. Some of the work is specifically designed to reinforce skills in mental arithmetic, and there is also work on using calculators efficiently.
	8.2 Large numbers and rounding	1	• To round large numbers	
	8.3 Significant figures	1	• To round to one or more significant figures	
	8.4 Standard form with large numbers	2	• To write a large number in standard form	You can introduce standard form as a powerful tool, which is widely used in science.
	8.5 Multiplying with numbers in standard form	1	• To multiply with numbers in standard form	
	Challenge - Space – to see where no one has seen before	1		This activity is designed to combine the skills covered across this chapter to explore an interesting real-life problem in a slightly more abstract context.
Chapter 6–8 assessment on Collins Connect				
9 Interpreting data	9.1 Pie charts	1	• To work out the sectors in pie charts by their angles at the centre	This chapter builds on previously learnt statistical principles. It extends pupils' use of data and knowledge of how to interpret statistical diagrams and charts. This is vital if pupils are to understand and interrogate the data being presented.
	9.2 Creating pie charts	1	• To use a scaling method to draw pie charts	
	9.3 Scatter graphs and correlation	1	• To read scatter graphs • To understand correlation	
	9.4 Creating scatter graphs	2	• To create scatter graphs	
	Challenge - Football attendances	2		This activity consolidates the previous work on statistics.
Half-term				

Half-term / Term 4				
10 Algebra	10.1 Algebraic notation	1	• To simplify algebraic expressions involving the four basic operations	Introduce algebra as a universal language with rules that are used all over the world. Discuss a range of examples in which algebra is used. Pupils often struggle to appreciate that letters represent variables and try to substitute particular values for the letters. Give pupils plenty of opportunity to reflect on the use of algebra as generalised number and to make clear links to the rules they have learnt for number.
	10.2 Like terms	1	• To simplify algebraic expressions by combining like terms	
	10.3 Expanding brackets	1	• To remove brackets from an expression	
	10.4 Using algebraic expressions	2	• To manipulate algebraic expressions • To identify equivalent expressions	
	10.5 Using index notation	2	• To write algebraic expressions involving powers	
	Mathematical reasoning – Writing in algebra	2		This activity develops confidence and fluency with algebraic notation. Pupils often struggle to decode everyday language into mathematics. This activity gives them the opportunity to practise this in a range of contexts.
11 Congruence and scaling	11.1 Congruent shapes	1	• To recognise congruent shapes	Pupils often do not realise that you can test for congruence by placing one shape on top of the other. Encourage the use of tracing paper to do this. Also reinforce the fact that shapes can have different orientations and still be congruent. Pupils can often use an incorrect point as the centre of enlargement or often just enlarge the shape without reference to the given point.
	11.2 Enlargements	1	• To enlarge a 2D shape by a scale factor	
	11.3 Shape and ratio	2	• To use ratio to compare lengths, areas and volumes of 2D and 3D shapes	
	11.4 Scales	1	• To understand and use scale drawings • To know how to use map ratios	
	Problem solving – Photographs	2		This activity consolidates topics previously covered on extracting data, area and ratio.
Chapter 9–11 assessment on Collins Connect				
Holidays				
Half-term / Term 5				
12 Fractions and decimals	12.1 Adding and subtracting fractions	2	• To add and subtract fractions and mixed numbers	Help pupils to understand the relationship between decimals and fractions as being different representations of parts of a whole.
	12.2 Multiplying fractions and integers	2	• To multiply a fraction and an integer	
	12.3 Dividing with integers and fractions	2	• To divide a fraction or a mixed number by an integer • To divide an integer by a unit fraction	
	12.4 Multiplication with large and small numbers	1	• To multiply with combinations of large and small numbers mentally	
	12.5 Division with large and small numbers	1	• To divide combinations of large and small numbers mentally	
	Challenge – Guesstimates	1		This activity gives pupils the opportunity to practice their mental strategies in some real-life contexts. It also encourages pupils to make links to the use of estimation as well as the need to make assumptions when tackling real-life problems.
13 Proportion	13.1 Direct proportion	1	• To understand the meaning of direct proportion • To find missing values in problems involving proportion	Pupils will often mix up direct and inverse proportion usually using direct proportion to answer inverse proportion questions.
	13.2 Graphs and direct proportion	1	• To represent direct proportion graphically and algebraically	

	13.3 Inverse proportion	1	• To understand what inverse proportion is • To use graphical and algebraic representations of inverse proportion	
	13.4 Comparing direct proportion and inverse proportion	1	• To recognise direct and inverse proportion and work out missing values	
	Challenge – Planning a trip	1		For this challenge pupils apply their understanding of proportion to a typical real-life context including speed, time and fuel consumption. The questions increase in complexity and pupils can use a range of graphical and algebraic skills to tackle them. They also need to be able to interpret some quite complex language.
Chapter 12–13 assessment on Collins Connect				
14 Circles	14.1 The circle and its parts	1	• To know the definition of a circle and the names of its parts	Pupil's often confuse radius and diameter. Give them plenty of opportunity to use both. Pupils often do not make the link between the work they have done previously on perimeter and area and the work on the circumference and area of a circle.
	14.2 Circumference of a circle	1	• To work out the relationship between the circumference and diameter of a circle	
	14.3 Formula for the circumference of a circle	1	• To calculate the circumference of a circle	
	14.4 Formula for the area of a circle	1	• To calculate the area of a circle	
	Financial skills – Athletics stadium	2		This activity is designed to give pupils the opportunity to apply their knowledge to a multi-step real-life problem. The context is common, but is presented in a slightly more complex way than pupils are used to.
			Half-term	
Half-term / Term 6				
15 Equations and formulae	15.1 Equations with brackets	1	• To solve equations involving brackets	A common problem often seen when expanding a bracket is to multiply the first term by the number outside the bracket and just write down the second term. Pupils will sometimes get confused with adding or subtracting from each side when dealing with equations with unknowns on both sides.
	15.2 Equations with the variable on both sides	1	• To solve equations with the variable on both sides	
	15.3 More complex questions	2	• To solve equations with fractional coefficients. • To solve equations with brackets and fractions	
	15.4 Rearranging formulae	1	• To change the subject of a formula	
	Mathematical reasoning – Using graphs to solve equations	1		In this activity pupils use mathematical reasoning to make links between equations and formula and their graphical representation. By comparing graphical and algebraic representations pupils check their ability to solve equations. This ability to use different representations to check their understanding is a valuable generic skill.
16 Comparing data	16.1 Grouped frequency tables	1	• To create a grouped frequency table from raw data	Encourage pupils to think about how statistics are used. Pupils need to consider how to present information.

	16.2 Drawing frequency diagrams	1	• To interpret frequency diagrams • To draw a frequency diagram from a grouped frequency table	Pupils also need to think about how we use statistics to model populations where it is difficult, or in many cases impossible, to gather all the population information.
	16.3 Comparing data	2	• To use mean and range to compare data from two sources	
	16.4 Which average to use?		• To understand when each different type of average is most useful	
	Problem solving – Technology questionnaire	1		This activity is designed to combine all the lessons in this chapter by taking pupils sequentially through the steps of tabulating and displaying data for a very familiar real-life problem. All the data is given, but pupils will need to read and think carefully about how they display the data so that they can make valid comparisons.
Chapter 14–16 assessment on Collins Connect				
End of year assessment on Collins Connect				

2-year scheme of work

The following scheme of work provides a suggestion for teaching Pupil Book 2.2 as part of a 2-year Key Stage 3 course.

Please note that you can recombine the test questions provided on Collins Connect to create new tests if your frequency of assessment differs from that below, or if you wish to combine content from different chapters in your own half-term tests.

This scheme of work is provided in editable Word and Excel format on the CD-ROM accompanying this Teacher Pack.

Chapter	Lesson	No. of hours	Learning objective	Comments/ suggestions
Half-term / Term 1				
1 Working with numbers	1.1 Multiplying and dividing negative numbers	1	• To carry out multiplications and divisions involving negative numbers	Much of this material with be new to Year 8 pupils. Pupils can leave out questions 1 and 2 of Exercise 1A, which was covered in Year 7. If pupils are grasping the concepts in this chapter they can move swiftly through the exercises, leaving out some of the questions.
	1.2 Factors and highest common factors (HCF)	1	• To understand and use highest common factors	
	1.3 Lowest common multiple (LCM)		• To understand and use lowest common multiples	
	1.4 Powers and roots	1	• To understand and use powers and roots	
	1.5 Prime factors	1	• To understand what prime numbers are • To find the prime numbers of an integer	
	Challenge – Blackpool Tower	1		This activity is designed to give pupils the opportunity to apply their learning to a real-life multi-step problem.
2 Geometry	2.1 Angles in parallel lines	1	• To calculate angles in parallel lines	Much of the material in this chapter will be familiar to learners. Use the activities and challenges at the end of each lesson to check understanding. If this is secure, move straight to Lesson 2.5.
	2.2 The geometric properties of quadrilaterals	1	• To know the geometric properties of quadrilaterals	
	2.3 Rotations	1	• To understand how to rotate a shape	
	2.4 Translations	1	• To understand how to translate a shape	
	2.5 Constructions	1	• To construct the mid-point and the perpendicular bisector of a line • To construct an angle bisector	
	Challenge – More constructions	1		This challenge gives pupils the opportunity to extend their learning to more complex constructions. They need to be able to reproduce a set of instructions that extend what they have already done in the lesson.
Chapters 1–2 assessment on Collins Connect				
3 Probability	3.1 Probability scales	1	• To use a probability scale to represent a chance	Much of this material will be new. Pupils may be familiar with Lesson 3.1 from Year 7 and can move to the activity question at the end if this is the case.
	3.2 Mutually exclusive events	1	• To recognise mutually exclusive events	
	3.3 Using a sample space to calculate probabilities	1	• To use sample spaces to calculate probabilities	
	3.4 Experimental probability	1	• To calculate probabilities from experiments	

	Financial skills – Fun in the fairground	1		In this activity learners extend their understanding of probability to a common real-life application that they may not have previously considered. This activity also makes a real-life link between probability and financial skills.
Half-term				
Half-term / Term 2				
4 Percentages	4.1 Calculating percentages	1	• To write one quantity as a percentage of another • To use percentages to compare quantities	Although pupils have met percentages before there are some important and quite challenging concepts in this chapter. The idea of percentages as a multiplier and the use of multiplicative reasoning is very important to pupils' confidence and fluency with percentages. Therefore, while you may be able to leave out some of the earlier questions in each section, be careful about leaving out too much or moving on too fast.
	4.2 Calculating percentage increases and decreases	1	• To use a multiplier to calculate a percentage change	
	4.3 Calculating a change as a percentage	1	• To work out a change in value as a percentage increase or decrease	
	Challenge – Changes in population	1		This activity is designed to give pupils the opportunity to demonstrate their understanding of percentage change to a real-life situation. All the information they need is provided but they will need to read the question carefully to decide which information they need and what mathematical skills to use.
5 Sequences	5.2 The *n*th term of a sequence	1	• To use the *n*th term of a sequence	Pupils can jump to the investigation on the *n*th term if they have met this in Year 7.
	5.3 Working out the *n*th term of a sequence	1	• To work out the *n*th term of a sequence	
	5.4 The Fibonacci sequence	1	• To know and understand the Fibonacci sequence	
	Investigation – Pond borders	1		Pupils apply their understanding of sequences to a real-life scenario. They will need to work methodically and be able to justify their solutions. Ask more able pupils to generalise their rules for an $m \times n$ pool.
Chapters 3–5 assessment on Collins Connect				
6 Area of 2D and 3D shapes	6.1 Area of a triangle 6.2 Area of a parallelogram	1	• To work out the area of a triangle • To work out the area of a parallelogram	Pupils should be familiar with many of the concepts in this chapter. Check their understanding with a couple of examples and the move to the MR, PS and challenge or investigation questions in each lesson. You may want to combine lessons 1 to 3.
	6.3 Area of a trapezium	1	• To work out the area of a trapezium	
	6.4 Surface areas of cubes and cuboids	1	• To find the surface areas of cubes and cuboids	
	Investigation – A cube investigation	2		Pupils apply their understanding of area to a more complex problem. They will need to work methodically and be able to explain their solutions. Ask more able pupils to justify any rules by revisiting the structure of the problem.
7 Graphs	7.1 Graphs from linear equations	1	• To recognise and draw the graph of a linear equation	It is important to take time over the examples in this chapter. However, it may often be more worthwhile to work through one or two examples in depth as a class, followed by picking out one or two examples for pupils to complete.
	7.2 Gradient (steepness) of a straight line	1	• To work out the gradient in a graph from a linear equation • To work out an equation of the form $y = mx + c$ from the graph	
	7.3 Graphs from simple quadratic equations	1	• To recognise and draw the graph from a simple quadratic equation	

	7.4 Real-life graphs	1	• To draw graphs from real-life situations to illustrate the relationship between two variables	
	Challenge – The M25	1		A common response to algebra is to ask how it can be used. This activity provides an everyday use of algebra in terms of graphical representation of algebraic relationships set in real-life contexts. Encourage pupils to suggest possible questions.
		Holidays		
Half-term / Term 3				
8 Simplifying numbers	8.1 Powers of 10	1	• How to multiply and divide by powers of 10	There are new ideas in all these lessons, which build on pupils' existing knowledge of rounding and the number system. Check understanding by doing a couple of examples as a class; then ask pupils to focus on the PS and MR questions, activities and investigations.
	8.2 Large numbers and rounding	1	• How to round large numbers	
	8.3 Significant figures	1	• To round to one or more significant figures	
	8.4 Standard form with large numbers	1	• To write a large number in standard form	
	8.5 Multiplying with numbers in standard form	1	• To multiply with numbers in standard form	
	Challenge - Space – to see where no one has seen before	1		This activity is designed to combine the skills covered across this chapter to explore an interesting real-life problem in a slightly more abstract context.
Chapters 6–8 assessment on Collins Connect				
9 Interpreting Data	9.1 Pie charts 9.2 Creating pie charts	1	• To work out the sectors in pie charts by their angles at the centre • To use a scaling method to draw pie charts	
	9.3 Scatter graphs and correlation	1	• To read scatter graphs • To understand correlation	Much of the material in lessons 9.3 and 9.4 will be new to pupils. However, the material could again be combined. Make certain that pupils have a good grasp of correlation before moving on.
	9.4 Creating scatter graphs	1	• To create scatter graphs	
	Challenge - Football attendances	2		This activity consolidates the previous work on statistics.
10 Algebra	10.1 Algebraic notation 10.2 Like terms	1	• To simplify algebraic expressions involving the four basic operations • To simplify algebraic expressions by combining like terms	Pupils should have met the concepts in lessons 1 and 2 before. Work through a couple of examples to check understanding and then move on to Lesson 3.
	10.3 Expanding brackets	1	• To remove brackets from an expression	
	10.4 Using algebraic expressions	1	• To manipulate algebraic expressions • To identify equivalent expressions	
	10.5 Using index notation	1	• To write algebraic expressions involving powers	
	Mathematical reasoning – Writing in algebra	2		This activity develops confidence and fluency with algebraic notation. Pupils often struggle to decode everyday language into mathematics. This activity gives them the opportunity to practise this in a range of contexts.
11 Congruence and scaling	11.1 Congruent shapes	1	• To recognise congruent shapes	Pupils will have met some of the basic concepts in this chapter. If they can demonstrate that they are confident and fluent with these basic concepts,
	11.2 Enlargements	1	• To enlarge a 2D shape by a scale factor	

	11.3 Shape and ratio	1	• To use ratio to compare lengths, areas and volumes of 2D and 3D shapes	pupils can move on to the more challenging questions at the end of each exercise in the Pupil Book.
	11.4 Scales	1	• To understand and use scale drawings • To know how to use map ratios	
	Problem solving – Photographs	2		This activity consolidates topics previously covered on extracting data, area and ratio.
Chapter 9–11 assessment on Collins Connect				
Half-term				
Half-term / Term 4				
12 Fractions and decimals	12.1 Adding and subtracting fractions	1	• To add and subtract fractions and mixed numbers	Much of the material in this chapter will be unfamiliar to pupils. Make sure that all pupils fully understand each concept before moving on to the MR and PS questions in the exercises in the Pupil Book.
	12.2 Multiplying fractions and integers	1	• To multiply a fraction and an integer	
	12.3 Dividing with integers and fractions	1	• To divide a fraction or a mixed number by an integer • To divide an integer by a unit fraction	
	12.4 Multiplication with large and small numbers	1	• To multiply with combinations of large and small numbers mentally	
	12.5 Division with large and small numbers	1	• To divide combinations of large and small numbers mentally	
	Challenge – Guesstimates	1		This activity gives pupils the opportunity to practise their mental strategies in some real-life contexts. It also encourages them to make links to the use of estimation as well as the need to make assumptions when tackling real-life problems.
13 Proportion	13.1 Direct proportion	1	• To understand the meaning of direct proportion • To find missing values in problems involving proportion	Much of this material in this chapter will be unfamiliar to pupils. Make sure that each concept is fully understood by all pupils before moving on to the MR and PS questions in the exercises.
	13.2 Graphs and direct proportion	1	• To represent direct proportion graphically and algebraically	
	13.3 Inverse proportion	1	• To understand what inverse proportion is • To use graphical and algebraic representations of inverse proportion	
	13.4 Comparing direct proportion and inverse proportion	1	• To recognise direct and inverse proportion and work out missing values	
	Challenge – Planning a trip	1		For this challenge pupils apply their understanding of proportion to a typical real-life context including speed, time and fuel consumption. The questions increase in complexity and pupils can use a range of graphical and algebraic skills to tackle them. They also need to be able to interpret some quite complex language.
Chapter 12–13 assessment on Collins Connect				

14 Circles	14.1 The circle and its parts 14.2 Circumference of a circle	1	• To know the definition of a circle and the names of its parts • How to work out the relationship between the circumference and diameter of a circle	Pupils may be familiar with the contents of lessons 1 and 2. Check understanding with a couple of example and if pupils are confident and fluent move straight on to lessons 3 and 4.
	14.3 Formula for the circumference of a circle	1	• To calculate the circumference of a circle	
	14.4 Formula for the area of a circle	1	• To calculate the area of a circle	
	Financial skills – Athletics stadium	2		This activity is designed to give pupils the opportunity to apply their knowledge to a multi-step real-life problem. The context is common, but is presented in a slightly more complex way than pupils are used to.
15 Equations and formulae	15.1 Equations with brackets 15.2 Equations with the variable on both sides	1	• To solve equations involving brackets • To solve equations with the variable on both sides	Much of this chapter will be new material. However, pupils who are familiar with multiplying brackets and solving simple equations can quickly complete Exercise 15A or move straight on to exercise15B.
	15.3 More complex questions	1	• To solve equations with fractional coefficients • To solve equations with brackets and fractions	
	15.4 Rearranging formulae	1	• To change the subject of a formula	
	Mathematical reasoning – Using graphs to solve equations	1		In this activity pupils use mathematical reasoning to make links between equations and formula and their graphical representation. By comparing graphical and algebraic representations pupils check their ability to solve equations. This ability to use different representations to check their understanding is a valuable generic skill.
16 Comparing data	16.1 Grouped frequency tables 16.2 Drawing frequency diagrams	1	• To create a grouped frequency table from raw data • To interpret frequency diagrams • To draw a frequency diagram from a grouped frequency table	Use one or two examples to check understanding from lessons 1 and 2, and if pupils are fluent and confident with the concepts, move straight to lessons 3 and 4. Compare and make decisions on the most appropriate statistical measures.
	16.3 Comparing data 16.4 Which average to use?	1	• To use mean and range to compare data from two sources • To understand when each different type of average is most useful	
	Problem solving – Technology questionnaire	1		This activity is designed to combine all the lessons in this chapter by taking pupils sequentially through the steps of tabulating and displaying data for a very familiar real-life problem. All the data is given but pupils will need to read and think carefully about how they display the data so that they can make valid comparisons.
Chapter 14–16 assessment on Collins Connect				
End of year assessment on Collins Connect				
Holidays				
Half-term / Term 5				
Work continues with Pupil Book 2.3				
Half-term				
Half-term / Term 6				
Work continues with Pupil Book 2.3				

Notes

Notes

Notes

Notes